# HUMANITARIANISM 2.0

HUGO SLIM

# Humanitarianism 2.0

*New Ethics for the Climate Emergency*

in association with

**Trócaire**
TOGETHER FOR A JUST WORLD

**BritishRedCross**

**German Red Cross**

**caritas**
Germany

**DANISH RED CROSS**

**CRS**
CATHOLIC RELIEF SERVICES

HURST & COMPANY, LONDON

First published in the United Kingdom in 2024 by
C. Hurst & Co. (Publishers) Ltd.,
New Wing, Somerset House, Strand, London, WC2R 1LA
© Hugo Slim, 2024
All rights reserved.

Distributed in the United States, Canada and Latin
America by Oxford University Press, 198 Madison Avenue,
New York, NY 10016, United States of America.

A Cataloguing-in-Publication data record for this book
is available from the British Library.

ISBN: 9781911723707

www.hurstpublishers.com

Printed in Great Britain by Bell & Bain Ltd, Glasgow

Just realize where you come from:
this is the essence of wisdom.
In harmony with the Tao,
the sky is clear and spacious,
the earth is solid and full,
all creatures flourish together,
content with the way they are.

Lao Tzu, *Tao Te Ching*

*For Liana, Maarten and Catherine-Lune*

# Contents

# Acknowledgements

A book is always a collaborative effort. I am particularly grateful to the three Red Cross organizations and three Caritas agencies who funded the research project that produced this book: Trocaire, Caritas Germany, Catholic Relief Services, German Red Cross, British Red Cross and Danish Red Cross. Together they formed an Advisory Group of committed people who were very kind to back me as I plunged into a new field of study. Noreen Gumbo, Ole Hengelbrock, Claire Clement, Peter Klanso, Gina Castillo and Christof Johnen formed the core group. For Christof, it was the second time he had generously trusted me to write a book. While the project was still only a thought, Jeanette Wijnants, Rasmus Stuhr Jakobsen, Louise Piel McKay, Hanne Mathisen, Andrea Steinke, Julie Arrighi, Catalina Jaime, Andrew Kruczkiewicz, Simon Addison, Greg Puley and Gerard Howe encouraged me to pursue the idea.

In Oxford people inspired me and gave a home to the project. I learnt a lot from several of Oxford's experts, including: Henry Shue, Simon Caney, Myles Allen, Lavanya Rajamani, Thomas Hale, Rupert Stuart-Smith, Bill Finnegan,

Erika De Berenguer Cesar, Richard Chen, Nathalie Seddon, Audrey Wagner, Carlota Segura Garcia, Alice Evatt and Asli Salihoglu. My research assistant, Mercedes Kuri, played a key role in getting me up to speed with the science and sociology of climate change, and Luca Marsico helped me to think about humanitarian needs. Many of these people have taught in climate workshops we have held for the Oxford Consortium for Human Rights where I have also learnt so much from young American students and enjoyed the unfailing support of Cheyney Ryan, Sujata Gadkar-Wilcox, Katie Dwyer, Tamara Niella, Ashleigh Landau, Alex Branzell, Susan Dichter, David Mwambari and Jessie DePonte and all the Consortium faculty. I am similarly grateful to students at Schwarzman College at Tsinghua University in Beijing where I have piloted my thinking and from whom I have learnt a lot, and to the Chinese Red Cross for inviting me to speak at the important Soochow International Humanitarian Forum on climate change in 2023.

Richard Finn, Anneli Chamblis Howes, Clare Broome Saunders, Kinga Rona-Gabnai, Barbara Brecht and Edward Hadas embraced the project at the Las Casas Institute at Blackfriars Hall where John O'Connor, Timothy Radcliffe and all the Friars warmly welcomed me and played their reassuringly mystical part in my work as usual.

Shiling Xu, Diana Ongiti, Jonathan Stone, Amanda Rives and Thorsten Gobel very kindly read and commented on a first draft, adding much value. The team at Hurst have been outstanding as always and a joy to work with, especially Michael, Mei, Lara, Daisy and Niamh.

At home I have been loved and encouraged by the support of friends and family. My sister, Mary Ann, and Herbie have been a constant source of encouragement. Jessie, Solly and Sophie have talked with me about climate and given me ideas. Susie, Isla, Winnie and Maisie have given me important experience of animal life, and Lilac kindly provided a home while ours was being rebuilt. Most of all, Asma's extraordinary love and joyful companionship have filled me with happiness each day.

Three Red Cross people have been especially important in the writing of this book. Maarten van Aalst brilliantly led the Red Cross and Red Crescent Climate Centre. He introduced me to the humanitarian challenge of climate change and suggested that I write a book. Catherine-Lune Grayson pioneered the ICRC's work on climate change and taught me a lot in the process. Liana Ghukasyan led the IFRC's humanitarian diplomacy at the UN in New York. She encouraged me to try out my ideas on states and agencies at the UN, and has supported and inspired me all along the way. This book is dedicated to all three of them with gratitude and admiration.

# Note on the Text

This book reflects the author's views alone and does not necessarily represent the positions of the members of the advisory group or the organizations that they represent.

# Preface

This book is about the humanitarian challenge of climate change. Its main purpose is to update humanitarian ethics for the era of climate emergency that presses so urgently upon humanity and nature. Humanitarian ethics and principles guide the work of helping people in extremis and regularly evolve to keep up with the times. Today is no exception. New ethics and new practices are vital as humanitarian workers and aid organizations adapt to a whole Earth emergency and a new scale of disasters in the late 2020s and 2030s, which will be driven by destructive extremes of heat, water, wind and fire on human life and nature caused by human-induced climate change.

I hope the book fulfils two tasks. Firstly, it is written in part as a basic introduction to the climate emergency for those who have not read deeply into the science and scholarship of climate change. I am aware that there are many humanitarians like me who have spent most of their career intensely focused on the humanitarian response to war and are only now looking at the humanitarian challenge posed by climate change. Various chapters and sections of the book are therefore intended to be a quick and easy way for them to get up to speed with the basic causes, dynamics and impact of the climate emergency.

Secondly, the main task of the book is to start the process of creating a whole new moral paradigm for humanitarian aid in the climate emergency. It is clear to me that the ethical framework we have inherited from the 1960s and 1990s is not fit for purpose in the climate emergency. We need to update it with a new understanding of what it means to be human and humanitarian when the very Earth on which we live is at risk of changing so significantly, and is already beginning to do so.

In setting out this new ethical framework for humanitarianism, I define the climate crisis as an Earth emergency and propose new doctrines of humanity and impartiality, which value the life of nature as well as human life, and also take more account of the future by embracing the principles of anticipation and adaptation. I also suggest that three core concepts of humanitarian practice—precaution, vulnerability and resilience—should be elevated to a firmer footing as key principles in humanitarian ethics, and that humanitarians should be especially mindful when working on mobility and loss.

The book is essentially a work of foundational ethics. It is not a practical discussion of operational ethics to guide people in the many difficult situations they face on the ground. Instead, the book thinks through the fundamental values we need as humanitarians today and tries to elaborate them more clearly as primary principles of purpose and action. Like many people, I sense that the climate emergency is an extremely important moment to revisit and renew our foundational ethics. I hope what follows is some help to the humanitarian profession at a time when it is reconsidering its purpose and approach.

The main argument of the book is that humanitarianism is not only for humans but for nature too. I hope this strikes a chord. I certainly sense this moral shift arising urgently within my own reason and conscience, and hear it from other humanitarians too. People in history and in traditional societies have obviously known this truth before, and most people around the world today are promoting it in earnest.

As always, I have written the book largely for humanitarians, but I hope it will be useful to people in other professions as well. I have also tried to write it accessibly by avoiding jargon and steering clear of obscure academic discourse. Please forgive me whenever I have failed in this effort. The text inevitably resonates with my own positionality. I imagine it is very easy to sense the kind of person I am but, if you have not guessed already, I will offer a brief description of myself.

I am a sixty-something white British male who has been educated in the classical European tradition in English private schools. I am originally and essentially a theologian from my undergraduate days, and my university teaching and humanitarian studies have taken me into international relations, moral philosophy and the general inter-disciplinary social thought of much Western academia today. I have worked as a humanitarian for several periods of my life in frontline operations, on policymaking and as a humanitarian diplomat. I have travelled widely, always carrying the baggage of my positionality with me as I go, and finding that most people can recognize it, tolerate it and work with it most of the time, and are reasonably polite when they find it too much at odds with their own experience of life.

Thank you for starting this book. I hope it proves useful however far you read.

*Oxford, Ascensiontide, 2024*

# 1

# Updating Humanitarianism

The climate emergency is bringing a new burst of creativity to the humanitarian sector. Billions of people vulnerable to climate-related disasters are struggling to find new ways to cope or adapt, and staff in humanitarian agencies, governments and community groups are discovering new techniques to help people survive better. An emergency is always an innovative time to be a humanitarian, and the climate emergency is both a global one and likely to run for decades to come.

What we call the climate emergency is, of course, a combination of different crises. It is primarily an *energy crisis*. Burning and processing fossil fuels, like coal, oil and gas, has enabled the development of the modern world. Carbon powered industry and technologies have brought many benefits and comforts to billions of people, but these fuels have cast a terrible shadow. Huge emissions of carbon dioxide, methane and other greenhouse gases have caused an *atmospheric crisis*. Our atmosphere, especially that precious 55-kilometre band of gases that cuddles our planet and makes it habitable for humans and millions of other species, is dangerously polluted, heating up and changing in ways that are unprecedented in the human experience of Earth.

This atmospheric change is creating a *weather crisis* by increasing the frequency and intensity of storms, floods, droughts and heatwaves. Changes in the weather are in turn creating an *environmental crisis* as landscapes change fast around us. Some areas are becoming much more threatening and less productive for humans. Others are opening up after millennia as permafrosted wilderness and, in the process, may tragically release even more carbon and water along with ancient viruses.

These critical changes in pollution, weather and environment are driving a *nature emergency*, which sees many forms of non-human life in crisis or in flux, and many finely balanced ecosystems changing or dying. The nature emergency is most commonly described as a *biodiversity crisis* in which various species of animals, plants, insects and microbes become increasingly rare, spread to new areas or surge in new environmental conditions.

Finally, of course, the combination of these energy, atmospheric, weather, environmental and nature emergencies is causing a *human emergency*. Our conditions of life, which have been relatively stable for many thousands of years, are changing fast. Where we live is getting hotter, wetter, windier, drier, often in strange combinations of all these. This is affecting the food we can grow, the water we rely on, the shelter we need and the illnesses we have. It is also diminishing our assets by damaging our homes, ending certain jobs and livelihoods, and sometimes making places where people have lived for centuries increasingly uninhabitable. As always, humans are working hard to adapt. The main challenge is to achieve positive and sustainable tipping points in renewable energy, climate resilience, protected ecosystems and new green jobs before nature's tipping points kick in to create a catastrophic and irreversible world, one in which human life and nature as we know it today are unable to survive or thrive.

The big change in all of this for humanitarians today is that we are faced by an *Earth system crisis*. This is different from occasional disasters that come and go, or terrible wars in which large groups of humans set out deliberately to destroy one another in particular places. The Earth crisis demands a new scale of thinking and a new ethics and consciousness in humanitarian work. This ethics must think about all life—not only human life—and must recognize survival as a joint project between humanity and nature. In the Earth crisis, humanitarians need to be more ecological and discover a humanitarian ethics that is fit for the Anthropocene, or the Ecocene, as more radical theorists are beginning to describe it better.

This process of ethical renewal is already well underway in the humanitarian sector and the wider field of climate action. Ethics always evolve to keep pace with new human experience, and humanitarian ethics are doing so now. Humanitarians have been leaning into environmental ethics for a long time, both in disaster relief and in legal protections of the environment in war. It is now time to recognize more clearly the vital connection between humanity and nature, and formalize it in a new set of moral commitments that is in step with the Earth emergency in which we live.

This book is intended to play a part in the necessary evolution of humanitarian ethics. At the outset, therefore, it is important to show how normal and important it is for us to update our ethics. Adapting our ethics and changing our principles for good reasons is not heretical or immoral. It is vital to register our constant awareness of what is right and wrong in a changing world around us. Without ethical innovation, humans would never have abolished child sacrifice and human slavery, or outlawed murder, torture and rape.

So, we must always be ready to change our moral values and the ethical codes with which we implement them when the world around us and our inner conscience presents us with a new sense of what is good and bad, right and wrong. Many humanitarians are deeply attached to today's humanitarian principles, but these too were the result of continuous innovation and adaptation, and now need to change again.

Equally, humanitarians do not exist in an ethical bubble or a closed system of emergency ethics which they alone decide. It is one of the great joys of life that humans connect and influence each other across borders of all kinds. Each one of us and each group of us are ethically porous. This means we are open to important ethical ideas from other cultures, other professions and other thinkers. The Western world's environmental reawakening is one of the great stories in the ethics of the last hundred years, and humanitarians are rightly influenced by it. In starting, therefore, it is also important to give a short history of the political evolution of environmental ethics in the last sixty years which shows how our humanitarian roots lie not only in the moral traditions of Geneva but also in ethical commitments made in Stockholm, Rio, Cochabamba, Sendai, Paris and Rome.

## *Evolving humanitarian ethics*

In October 2015, I went to a wonderful party in Vienna. I was one of a few lucky people at the International Committee of the Red Cross (ICRC) who were invited to celebrate the fiftieth anniversary of the Fundamental Principles of the Red Cross and Red Crescent Movement which had been agreed in Vienna in October 1965. Kindly hosted at the spectacular Italian Embassy (which was once the palace of Count Klemens von Metternich, Austria's great nineteenth-century statesman), the Italian Ambassador and the Austrian Red Cross gave us a memorable night. We listened to Red Cross and Red Crescent people from around the world telling seven stories about the seven principles in action—humanity, impartiality, neutrality, independence, voluntary service, unity and universality—and then we dined and danced.

Back in October 1965, the Twentieth International Conference of the Red Cross had met in even greater splendour at the Hofburg Palace in Vienna, the seat of the former Hapsburg Empire which ruled large parts of Europe from the late thirteenth century until its dramatic collapse in 1918. At the conference in 1965, 580 representatives of 92 national Red Cross and Red Crescent societies and 84 states had formally adopted the *Declaration of Red Cross Principles*. This confirmed seven Fundamental Principles as "a universal doctrine, a humanitarian basis common to all people", and agreed that the principles should be solemnly read aloud at the opening of every future conference.[1] Thus was established both the doctrine, and the indoctrination, of these principles, which continue to guide the movement today. Since the 1990s, the first four principles—humanity, impartiality, neutrality and independence—have also been taken up as ethical doctrine by United Nations agencies and many non-governmental organizations (NGOs).

The new Declaration of Principles in 1965 was a deliberate ethical innovation. Marina Sharpe's excellent history of the evolution of humanitarian principles describes how at an earlier meeting in Oxford in 1946, the Red Cross had agreed to be guided by four main principles, thirteen additional principles and six rules. One can rely on Oxford to complicate things. This cumbersome ethical system for Red Cross humanitarians lasted throughout the 1950s and early 1960s. Quite rightly, it seemed unnecessarily clunky to a brilliant young Swiss lawyer at the ICRC called Jean Pictet, who set about simplifying the principles in his PhD thesis. The ICRC then pushed Pictet's new seven principles through the wider Red Cross Movement and onwards to universal endorsement in Vienna.[2]

This historical vignette about humanitarian principles shows how values, norms and principles are always in motion, with continuity and innovation regularly complementing one another as we confront new human challenges and discover new moral aspects of ourselves. Some values stay the same but are reframed in more contemporary language. But new principles are also discovered, or re-discovered, as vital to the times. Fresh moral insights deliver genuine ethical innovation. For example, Sharpe very interestingly shows how today's first principle of humanity was not even on the list of the four main principles in Oxford in 1946. Humanity was more assumed than professed. Instead, impartiality, independence, and the equality and unity of Red Cross Societies formed the inner core. It was one of Pictet's most important ethical innovations to spell out humanity and rank it first in his sleek new sixties schema.

Humanitarian ethics continued to develop. In 1994, the *Code of Conduct for Disaster Relief* was agreed and deliberately emphasized people's participation, community empowerment, sustainability and the use of non-racist imagery in disaster marketing.[3] These principles might have been moral blind spots to some humanitarians of the 1950s but were essential to 1990s thinking.

Next, it was moral concerns for programme quality, accountability and individual human rights that drove the ethical innovation in humanitarian standards. These new moral concerns characterized the development of the *Humanitarian Charter*, the *Sphere Standards* and the *Core Humanitarian Standard* in the 2000s.[4] Most recently, in 2021, the climate emergency has prompted a new set of seven commitments in the *Climate and Environment Charter for Humanitarian Organizations*.[5] This latest moral list rightly puts ethical emphasis on adaptation, environmental sustainability, community-led action, collaboration, influence and green accountability.

In this latest ethical innovation, the Red Cross and Red Crescent Climate Centre in The Hague has played an extremely creative and important role as an ethical entrepreneur within the humanitarian sector. Its brilliant team of scientists and humanitarians explained the complexities of climate science to humanitarians in simple ways and sought out the grounded connections between climate change, human need and humanitarian response in countries all around the world. It has then helped humanitarian agencies to invest in operational strategies of "climate smart" mitigation, adaptation and resilience, while also promoting these policies and practices diplomatically with governments and international organizations.[6] Indeed, the Climate Centre deserves a medal for pulling the humanitarian sector towards greater climate consciousness.

So, humanitarian ethics has been steadily evolving and is now entering the climate era with some important moral adjustments already in place. It has not been doing this in isolation. Indeed, the humanitarian moral community is a relatively small one amidst the much wider community of global ethics. Another much broader ethical evolution has been unfolding fast alongside conventional humanitarian ethics. This is the field of environmental and climate ethics, which has rightly had a significant influence over humanitarians working in climate-related disasters.

## The influence of environmental ethics

While Pictet was streamlining humanitarian ethics in the 1960s and expounding the principle of humanity in greater detail, largely for situations of war, other norm entrepreneurs were engaged in ethical renewal of a different kind. Many people around the world were becoming swept up in a much bigger vision of humanity's ethical challenge, which concerned the planet itself and not just human conflict in particular parts of the planet. A new sense of conscience and concern was brewing around humanity's relationship with nature. This began to be discussed in the new discourse of "environment".[7]

After the mass expansion of industrial technology and its use in the destructive power of modern warfare, people in Europe and North America began to recover a sense of the importance of nature, especially humanity's dependence upon and existence within it. Western environmentalists began to find common cause with millions of indigenous and impoverished people around the world whose environments were being badly degraded, and with Asian, African, Pacific and Native American thinkers who had always kept hold of the sanctity of the natural world and the essential embeddedness of humanity within it.

In 1962, three years before the Red Cross conference in Vienna, and after years spent studying and writing beautifully about the sea, the great American biologist Rachel Carson turned her attention to the danger of pesticides and pollution. Her book *Silent Spring* was a turning point in the modern environmental movement of the West and argued for greater respect for nature and a better balance between technological progress and the conservation of the environment. In 1972, British and

French environmentalists, Barbara Ward and Rene Dubos, wrote a similarly pioneering book called *Only One Earth: The Care and Maintenance of a Small Planet*. Ward then founded the International Institute for Environment and Development in London (IIED), which has played a key role in global environmental policymaking and still does. Years later in 2006, former US Vice President Al Gore would have a powerful and unexpected influence. His film *An Inconvenient Truth* brought climate science to popular attention around the world and energized the movement for climate action.

The "Green Movement" grew fast in the sixties, seventies and eighties, gradually attracting the attention of the political mainstream.[8] A significant part of the Green Movement converged around an almost mystical commitment to "deep ecology". This school of thought rejected the ancient and often patriarchal ideology of humans' supreme position over nature and our right to dominate it. Instead, rejecting "human exceptionalism", deep ecologists recognize humans as simply another part of nature which must respect its law and live modestly within its limits, having no greater priority than other forms of life.

Scandinavian governments championed the new global ethic of environmentalism bubbling up around the world and became politically associated with it in the 1970s and 1980s. A UN Conference on the Human Environment in Stockholm in 1972 produced a strong declaration which enshrined many of the ethical ideas that still shape environmental ethics and climate justice today. This recognized the finite limits of the Earth and its resources, and the need for development to take fairly into account Earth's present population and "a solemn responsibility" for future generations. It also firmly linked poverty, human rights and the environment in its very first principle, albeit in non-inclusive language, by affirming: "Man has the fundamental right to freedom, equality and adequate conditions of life, in an environment of a quality that permits a life of dignity and well-being".[9] The Stockholm conference set up the United Nations Environment Programme (UNEP) in the same year.

In 1983, the UN sponsored the former Norwegian Prime Minister, Gro Harlem Brundtland, as the first head of the World Commission for Environment and Development, to produce a report on the environment and development. Published in 1987, the Brundtland Report examined problems of global warming, ozone depletion, environmental degradation, food security, cities and the importance of ecosystems. Famously, it introduced the moral goal of "sustainable development" which it defined as "development that meets the needs of the present without compromising the ability of future generations to meet their own needs."[10] This new policy formalized the ethical commitment to environmental care and intergenerational responsibility that is now so central today.

The Brundtland Report set up the UN Commission on Sustainable Development, which then organized the genuinely world-changing Earth Summit in Rio de Janeiro in 1992. This was an exuberant multi-stakeholder summit (states and civil society) that was full of optimism and brought the worldview of indigenous people to the centre of international politics. Rio rode the growing wave of environmental awareness and enjoyed an easier political consensus because of the end of the Cold War.

The summit produced a *Declaration on Environment and Development*, the first article of which affirmed that human beings "are entitled to a healthy and productive life in harmony with nature." Rio also agreed an international treaty—the *UN Framework Convention on Climate Change* (UNFCCC)—which began the series of "conferences of the parties" to the treaty, or COPs, which are now at the centre of global policy on climate action. It also agreed Agenda 21, which inspired local governments all over the world to embrace environmentalism and launch green policies.

The UNFCCC defined many terms that we now know as integral to climate policy, such as climate change, emissions and carbon sinks. It introduced core policy concepts like mitigation, adaptation and the special vulnerability of developing countries and small island states. It also coined the important principle of "common but differentiated responsibilities and respective capabilities" (CBDRRC), which asserts that while all states must cooperate in climate and environmental action, some States have more responsibility for the crisis and greater capacity to pay for it than others. Therefore, the UNFCCC agreed the need for funding by developed countries and introduced key COP processes like national plans and reporting on climate action. Since Rio, the Inter-Governmental Panel on Climate Change (IPCC) has also played a central role around the UNFCCC. Set up by the World Meteorological Organization (WMO) and UNEP in 1988, the IPCC has consistently received mandates and instructions from the UN General Assembly and the COP, becoming the body of experts most routinely recognized and referred to by states as the basis for sound climate policy.

More than just the UNFCCC, the Rio Earth Summit also saw states agree two other important treaties on environmental protection: one to counter the threat of desertification, and the UN Convention on Biological Diversity (UNCBD) to conserve nature and ecosystems and ensure the fair use of natural resources. The UNBCD produced "the other COP". Its conference of the parties meets every two years on biodiversity, and had its COVID-delayed Kunming-Montreal COP 15 in 2023. Just as the UNFCC is advised by the IPCC, the UNCBD is advised by IPBES, which is the Intergovernmental Science-Policy Platform on Biodiversity and Ecosystem Services headquartered in Bonn. This too is an impressive intellectual powerhouse of researchers.

The importance of scientific research and ongoing study of climate change and biodiversity was strongly affirmed in the UNFCCC and the UNBCD, which are both determined to be evidence-based. Knowledge is understood as an essential virtue to guide policy and action on climate and biodiversity and help bring about both treaties' ethical ambition to stabilize climate change and conserve biodiversity for the common good of present and future generations.

It was COP 21 in Paris in 2015 which proved to be the next real game-changer in global climate ethics. The so-called Paris Agreement set out a mitigation target to limit average global temperature to a 1.5% increase above pre-industrial temperatures. It also spelled out the concepts of adaptation, loss and damage in more detail and required all countries to submit nationally determined contributions (NDCs) every five years, setting out how they plan to reduce emissions and increase adaptation and climate resilience.

While governments and civil society were preparing for the COP in Paris in early 2015, another global norm entrepreneur was polishing up a major policy paper for publication. Pope Francis was in Rome editing his new Catholic doctrine on the climate and environment emergency, which he published in May 2015. He called his Encyclical *Laudato Si'* after the opening words of the Canticle to the Sun by his medieval namesake and nature lover St Francis of Assisi. Religions have always been a rich source of environmental ethics. Yet surprisingly perhaps, in this modern age, it is the theological and moral innovations of Pope Francis that have had a significant political influence on the ethics of many people in civil society, humanitarian agencies and governments around the world.

Francis's two immediate predecessors, John Paul II and Benedict XVI, both developed Catholic environmental doctrine by warning against the destruction of the environment and its further impoverishment of poor people by technological hubris and capitalist greed. In 2002, John Paul II and the Orthodox Church's Patriarch Bartholomew I called for much greater "care for creation" and an increase in "ecological awareness".[11] In his inaugural homily in 2005, Pope Benedict lamented that "the Earth's treasures no longer serve to build God's garden for all to live in, but they have been made to serve the powers of exploitation and destruction". Benedict went on to install solar panels at the Vatican and had the Holy See invest in sustainable forests.

But it is Pope Francis who has set out a more complete doctrine on the climate and environment emergency in *Laudato Si'*. Building on the liberation theology of Brazilian priest, Leonardo Boff, from the 1990s, which combined the "cry of the Earth" with "the cry of the poor" in simultaneous commitments to ecology and social justice, Francis notes how environmental catastrophe is intrinsically linked to human suffering and poverty. Therefore, *Laudato Si'* demands an "integral ecology" which treats climatic, environmental, social and economic problems as one in a comprehensive quest for justice and peace.

An essential ingredient in the success of Francis' approach is an "ecological conversion" in every human heart that makes us see nature differently as valuable, beautiful and precious, and the Earth as our "common home". Most significantly, Francis stepped away from traditional Catholic doctrine that sees humans as a sort of master species with rights to dominate and rule over animals and plants. Instead, his new ecological theology distinguishes humanity simply as "unique" among creatures but embedded within nature like all other creatures. He also emphasizes a persistent Christian hope that all creation should be made perfect in heaven, not just humankind.

The refreshing spiritual call of Pope Francis clearly touched the hearts and minds of billions of people round the world. His ethics chime with environmental beliefs in all world religions and get close to the embedded worldview of many indigenous peoples. By connecting globally and speaking so passionately and prophetically about the climate emergency, Francis comes across as genuinely ecumenical and able to be heard by people of all faiths and none. With his voice and his message, he has shaped a global groundswell of ecological concern for climate action, environmental protection and biodiversity.

## *The rise of disaster ethics*

In the decades after Rio, climate-related disasters kept occurring and intensifying. Slowly, people began to link them with climate change. Humanitarian experience in these disasters fused with the emergence of environmental ethics to produce a new ethics and policy framework of disaster risk reduction (DRR). This prioritized the values of environmental protection alongside the practical virtues of anticipation, preparedness, mitigation and adaptation to reduce the destruction of human life, livelihoods, assets and nature. This approach to disaster ethics is now encapsulated in the meta-value of resilience, which is an ideal state that allows human and non-human life to withstand the shock of climate change and its fierce weather hazards.

The Japanese have become the global political champions of DRR and have hosted the World Conference of Disaster Reduction in the cities of Yokohama, Hyogo, Kobe and Sendai since 1994, helping to create the United Nations Office for Disaster Risk Reduction (UNDRR) in 1998. Today, the Sendai Framework agreed by States articulates the policy ambition and good practice of DRR. The International Federation of the Red Cross and Red Crescent Societies (IFRC) has been a long-time policy champion and expert proponent of DRR. So too have many Asian and Latin American NGOs who have been pioneers in locally led DRR, often as innovators and first responders to climate-related disasters.

Like the global ethics emerging from Rio and Rome, Sendai's ethics is seized by a commitment to find harmony between humans and nature. The great ethical discovery of disaster studies and the Sendai process has been to stop demonizing nature in the idea of "natural disasters" and instead recognize the human role in creating disasters by degrading the environment and tolerating poverty and bad spatial planning which renders billions of people vulnerable to water, wind, heat and cold. The moral insight of Sendai ethics is that humans and nature share responsibility for climate-related disasters. Both are, therefore, part of the solution. Humans must reduce their vulnerability and should work with nature to better protect themselves in nature-based solutions (NbS).

## Greening humanitarian ethics

Humanitarian values and principles cannot and should not remain untouched by these various waves of environmental ethics spreading through human minds, political agreements, international agencies and local communities as the world faces increasing floods, heatwaves, storms and droughts. While the human violence of war remains a major driver of human suffering and humanitarian needs, the violence of natural hazards in a warming world poses huge threats to the quality and conditions of human life. This is especially true when the world's human population is so big and so many live in densely populated towns and cities, with a significant majority in already precarious conditions with little protection against extremes of water, heat, cold and wind.

The nature of our times means humanitarian ethics and humanitarian action must become more ecological as human society tries to strike a better balance in its relationship with the natural world around us. This is the core challenge of updating humanitarian ethics and there is a lovely Italian word for it. Back in 1959, while Jean Pictet was making the final edits to his huge legal commentary on the Geneva Conventions, Pope John XXIII was launching Vatican II, the great Council of the Church which would modernize Catholicism and bring it firmly into the liberal age. Announcing his plan to convene the Council, he used the word *aggiornamento* to describe his purpose of bringing the Catholic Church up to date and making it relevant to the present time: "todaying" the Church. I hope this book helps the process of aggiornamento for humanitarian ethics in the 2020s.

In the chapters that follow, I will suggest important ways in which we can adapt our current humanitarian principles and innovate new ones so that they combine greater ecological and environmental purpose with their concern for human life. Before this, however, we must look at how climate change is creating a new era of emergency and examine the particular character of the climate emergency. Only when we have understood the kind of emergency we are in can we revise our ethics accordingly.

# 2

## An Earth Emergency

The climate emergency is an Earth emergency. It is urgent that we reduce its effects and adapt to its changes. I am using the term emergency very deliberately when talking about climate change. This is not because I am especially alarmist or apocalyptic about the future; I think humans will find an ecologically sensitive way through the changing climate both by limiting warming and adapting to it. We will reorganize ourselves as a species and continue. But I am not an easy optimist. Hundreds of millions of people, probably billions, will not be able to cope and adapt. Nor am I an easy utilitarian who thinks the main thing is that most of us get through it, while regretfully accepting that millions will not. The many people who will suffer in this long transition are rightly people and places of concern to humanitarians.

Emergency humanitarian response must play a significant part in climate action to support those who suffer most because they face extreme impoverishment or death from climate-related disasters. My ultimate optimism does not prevent me from recognizing climate change as an immediate global emergency. Indeed, my optimism is dependent on governments recognizing an emergency now—today—and thus acting with great urgency to reduce emissions, adapt fast and improve humanitarian techniques best suited to climate-related disasters.

Data is the tool we use to envisage the scale of the crisis now and in the future for human populations. The data we have for the future are inevitably estimates. The methods of collection and assessment are evolving as the collective intelligence of global human society attempts to understand the climate challenge. But a clear trend is already taking shape that confirms a global increase in people's experience of climate shocks and detects a significant percentage of people at high risk from such shocks. This all suggests that humanitarians can expect a very large and continuous surge in human suffering and impoverishment from climate change.

A 2023 study by Miki Khanh Doan and others at the World Bank has tried to assess the numbers of people who are exposed to, vulnerable to and at high risk from climate shocks.[1] They estimate that some 4.5 billion people in the world were exposed to extreme weather events in 2019, up from 4 billion in 2010. Of these, 2.3 billion people were on a low income and 400 million people lived in extreme poverty. Looking at people's vulnerability to climate shocks across a sample of 75 countries where data exists, the World Bank team estimate that 20% of all people exposed were at high risk from extreme weather. In some countries, people's risk has gone down despite increasing exposure. In poorer countries, however, this risk has gone up.

The Global Adaptation Initiative team (GAIN) at Notre Dame University measures countries' exposure to climate shocks and readiness to adapt to them to produce an all-country index. Overall, GAIN's data tracking shows that "people living in the least developed countries have 10 times more chance of being affected by a climate disaster than those in wealthy countries each year [and] it will take over a hundred years for lower income countries to reach the resiliency of richer countries."[2]

This, and similar analysis from the IPCC, all indicates that environmental, economic and social worlds will end for hundreds of millions of people in the next twenty years because of climate change. Their landscapes will become uninhabitable, their livelihoods will become impossible, and the conditions in which they live will become unbearable. Many will die, in particular from extreme heat and drought, and many others will be greatly impoverished by recurring damage to their assets, earnings and infrastructure. Even in places where people's way of life is not severely damaged, life will become more difficult as nature's tipping points gather speed and our modes of adapting may not guarantee the same quality of life. Maladaptation will also see people suffer from false starts and wrong turns in adaptation that set them on the pathway to failing change. Ecological worlds will also end for animals, fish, plants and insects, as they often have in ecological history, but this time we will see and understand the tragedy of their extinction. We will feel the loss of non-human friends and much of the essential mutualism between species on which all life depends, including our own.

Knowing that this is already happening in many places, and will intensify in the next few decades, makes climate change an emergency today. By definition, an emergency is something which is emerging before our eyes and demands urgent action from us. It makes no difference whether this rising danger is due tomorrow, next month or in ten years. Its impending severity requires that we act urgently now. Disasters do not have to be happening right now for their prevention to demand an urgent response today. A country does not wait to be invaded by an advancing enemy before it declares an emergency and takes urgent action to build up its defences. The real prospect of a disastrous future determines that the situation is an emergency already. This is especially true when the actions necessary to respond effectively to the emergency when it hits can only be taken in advance before it hits, like developing new technology, transforming organizations, training people, researching threats, building capabilities, and stocking up on reserves. None of this can wait until the day itself. The emergency starts now before complete disaster strikes.

## *The particular nature of climate risk*

Certain characteristics of climate change are already well understood, largely because of the accumulated work of the IPCC. Three words in particular keep popping up in the reports of the IPCC which reveal the distinct character of the climate emergency: irreversible, unprecedented and cascading.

The IPCC use *irreversible* to describe changes that have already happened which we cannot change, like temperature rises that have already taken place, ice that has already melted, water systems that have already dried up, and species that have already become extinct. Even if there is good progress in climate action in the next few years, climate scientists describe these irreversible changes as "residue" effects, which will stay with us to harm particular places and people. Therefore, heatwaves, sea changes and water shortages are already baked into our current world and will continue to have humanitarian consequences until everyone is well adapted to them. The IPCC points out that many other things will also be "irreversible" soon if we fail to reduce our emissions.

The second important word in IPCC reports for humanitarians is *unprecedented*. The unprecedented nature of future hazards and disasters means that, as humanitarians, we cannot predict the future from the past as we tend to do in emergency planning, preparedness and response. Things will happen which we have never seen before, like massive global needs and continuously recurring and intensifying hazards. Humanitarian need and response will not simply be more of the same because climate science suggests that disasters will soon be of a nature, scale, frequency and recurrence that we have never faced before.

One of the reasons for expecting climate-related disasters to be unprecedented is because of the third IPCC word— *cascading*. All disaster specialists recognize that a single climate-related factor, like transient extreme heat or a single hurricane, can trigger a chain of other disastrous conditions. For example, extreme heat in one of the world's bread baskets could seriously damage a geostrategic zone of food crops. The subsequent failure of supply into the global food market may then combine with existing inflation in other parts of the world to raise food prices beyond the reach of millions of people far away. Similarly, intense flooding could disable strategic electricity stations, with cascading effects on water pumping, industry, banking services, hospitals and food cooling systems. The rising heat in Brazil in 2023 and 2024 caused an exponential increase in cases of Dengue across the country.[3] Sometimes cascading effects can be predicted but at other times they cannot, which makes them especially risky. Reducing the risk of the initial climate-related trigger is the best way to lower the risk of an exponential cascade that reaches well beyond the initial disaster.

Beyond IPCC jargon, two other words also help to describe the climate emergency. Climate change is an emergency not simply because heatwaves are hotter, droughts are longer, floods are bigger and storms are stronger. It is an unusual emergency because these things will continue to happen everywhere and affect everyone. Another key characteristic of the new era of emergency, therefore, is that it is a *universal* emergency. IPCC reports have every country, every continent, every ocean and the whole atmosphere in view. The poor world and the rich world, north and south, black, brown and white, old and young, everyone and everywhere will experience climate emergency to a greater and lesser degree in the years ahead.

The universality of emergency is happening already. Just as we see huge floods in Pakistan, extreme heat in Delhi and intense cold in China, so too do we see large floods across Europe and huge fires across Canada, the USA and Australia. Climate change is an unusual emergency because, like a pandemic, it is making urgent demands for emergency aid all across the globe, and will continue to do so. Responding to emergencies everywhere at the same time poses a new challenge for humanitarian agencies, which will result in domestic humanitarian demands competing with international claims. Humanitarian response will be very expensive, so we can expect a crisis in emergency budgets to be a core feature of the climate crisis.

Finally, as is clear already, a distinctive characteristic of the climate emergency is that it is a *long* emergency. This longevity means that it makes sense to talk of a new era of emergency and recognize that climate change is not an episode of weeks or months but an emergency that will span several generations. In the next three decades, humanity will face a single long emergency as climate-related disasters follow seamlessly one upon another in different parts of the world. Many climate-related hazards naturally occur seasonally, like storms and floods, or they happen regularly as oscillations, like El Niño and La Niña, but we will see more of them everywhere, and often at surprising times. We will no longer experience climate-related disasters as episodic with long gaps between them, but as a continuous series of hazards. The accumulation of so many hazards and disasters makes it logical to talk of one long emergency, and orientate our mindset, planning and response to the idea of a humanitarian marathon rather than a series of sprints.

These five words—irreversible, unprecedented, cascading, universal and long—make it wise to recognize that we are in a new era of emergency.

## An earth systems emergency

The best way to understand and characterize this new era of emergency is as a whole Earth emergency. Many scientists are working hard along these lines and developing illustrative models and metaphors to communicate important data about changes to the Earth because of climate change.

Thinking in Earth terms is a significant change in moral scale and ethical concern for humanitarians. The great achievement of the humanitarian movement in the modern era has been to establish ethical and emotional recognition of the whole human species as a single object of moral concern. This is clearly expressed in the humanitarian idea that it is our "common humanity" which identifies and binds human beings everywhere as politically related and morally valuable. This species-wide recognition makes it important that we care for and care about all human beings as rightful claimants of compassion, help and justice. This is the ethical assumption of intrinsic human value that underwrites humanitarianism's first two principles of humanity and impartiality.

But the climate emergency, with its environmental and ecological crisis, demands a further evolution in our moral thinking as humanitarians. It calls us to feel for, recognize and respond to the Earth as common to us too. As humans, we are Earthlings. Indeed, the original Latin words for human and humanity have their root in the the word *humus* for earth. Humanity is of the Earth. The fact that humans are Earthlings makes Earth systems morally important to humanitarians.[4] As we will see in chapters 5 and 6, with the Earth system in crisis, it makes no moral sense to show ethical commitments only to humanity and not to the Earth.

So how exactly is our Earth, our home, in an ontological crisis because of climate change? Quite simply, if global warming and intense environmental change continue on their current path, the Earth is at risk of changing how it is. This is an ontological crisis because how the Earth is, and how it has been for millennia, suits how we are as humans. We are alive, and as we are, because the Earth is as it is. Changing the Earth will be disastrous for humanity and many other forms of life, for example wheat and rice, with whom we exist in various forms of mutual relationship.

The core ideology of modern humanitarianism has come of age alongside the rapid development of the humanities and social sciences that study all aspects of human life, like sociology, anthropology, psychology, philosophy, politics and law. Many of its practices have then deployed the medical, nutritional, economic and agricultural sciences. The ideology and ethics of today's Earth emergency is being shaped in an age of Earth sciences, like geology, meteorology, climatology, oceanography, ecology, environmental science and geography.

This Earth turn in science is rightly shifting human attention from who we are and how we should relate to one another to where we live. In so doing, Earth sciences are discovering exciting new knowledge and information about the Earth and extending humanity's ethical concern from people to planet. As we learn more about the Earth, we find good reasons to care more about it too. Greater scientific and ethical understanding arising from Earth sciences is inevitably and rightly changing humanitarian ethics and humanitarian action to move beyond a simple focus on the human.

Earth systems scientists use various metaphors and models to track the Earth emergency and describe its particular challenges and possible solutions. Three of these seem especially useful to help humanitarian understandings of the Earth systems: *boundaries, tipping points* and *Gaia*.

Johan Rockstrom, Joyeeta Gupta and many others at the Earth Commission have identified a range of "safe and just" *Earth systems boundaries* (ESBs) within which humans and nature can live fairly and well. Keeping within these boundaries will allow life to balance "Earth system resilience and human well-being." Inside these boundaries the Earth stays stable and is best for life.[5] In their model, Rockstrom and Gupta show that Earth safety can be maintained by keeping the Earth's various systems—the atmosphere, hydrosphere, geosphere, biosphere, cryosphere and pedosphere, or air, water, land, life, ice and soil—within certain limits that are optimal as a "life support system" for humans and nature.

Going further still, they argue ethically that Earth safety on its own is not enough. It must be complemented by "Earth system justice" too. Earth boundaries must be fair as well as safe. They suggest three measures of Earth justice which they call the 3Is: interspecies justice, intergenerational justice and intragenerational justice, which are also widely promoted by many other climate ethicists. Keeping within these boundaries involves living in a healthy "corridor". This is not a great metaphor because none of us really want to live in a corridor, but the point is a good one. Life of all kinds has always thrived within certain narrow ecological limits, as I was reminded 5000 metres high at the top of Mount Kenya last year, where I was remarkably less effective than I was a few hundred metres downhill. I would have been even worse a mere five metres under the sea. These boundaries and measures of justice make good moral goals for humanitarians and we will return to them in the chapters that follow.

If human activity bursts Earth system boundaries, we run the risk of hitting *harmful tipping points*. These are the second useful model to inform humanitarian thinking, mainly because they pose a worst-case scenario which must be avoided. The recent report on Global Tipping Points led by Tim Lenton at the University of Exeter, released at COP 28 in 2023, makes this point up front: "Harmful tipping points in the natural world pose some of the gravest threats faced by humanity. Their triggering will severely damage our planet's life-support systems and threaten the stability of our societies." Earth science shows us that climate change and nature loss together could cause abrupt and irreversible tipping points in which various Earth systems would fall out of their currently stable state and never recover, causing cascading and disastrous effects of all kinds "which could far exceed the ability of some countries to adapt".[6]

The 2023 report identifies a total of twenty-five Earth system tipping points. Six are in the icy cryosphere, like changes to the icesheets in Greenland and the Antarctic. Sixteen tipping points are in the biosphere, such as forest dieback in the Amazon and dryland degradation. Four are in the oceans and atmosphere circulations, like the West African monsoon. Already, five major tipping systems are at risk of crossing dangerous threshold levels even at the current level of global warming. These tipping points are no longer gauged as high-impact, low likelihood events but "are rapidly becoming high impact, high likelihood events." Avoiding such harmful tipping points demands a deep precautionary ethic in humanitarian action as it works with people around the world in different parts of the Earth system, and in humanitarian advocacy towards governments and businesses who have a much greater environmental impact.

The good news is that there are also *positive tipping points* and clear opportunities to tip them. These include the shift to renewable energy to reduce emissions and a green economy to create new livelihoods. Significant ecosystem regeneration is another. Achieving political tipping points in climate governance, cooperation and changed mindsets is enormously important. And, like harmful tipping points, positive tipping points can have cascading effects that produce a domino effect of increasing environmental goods. This makes it important to look for "super-leverage points" where small or large pressure could trigger an exponentially positive tipping point. For example, transforming maritime and airline fuels, or developing green ammonia, could all trigger positive tipping points. Humanitarian advocacy must join in the call for their prioritization.

The idea of *Gaia* uses another Earth system metaphor— the Greek goddess of the Earth—to encapsulate these two ideas of boundaries and tipping points in the image of Earth as a female and self-regulating system with innumerable feedback loops.[7] The theory of Gaia is a bit like scientific poetry and was created by James Lovelock, a distinguished British scientist and inventor of scientific instruments for NASA, and Lynn Margulis, the eminent American microbiologist. Working with their Earth system approach from the 1970s onwards and steeping it in the data of sophisticated biology and physics, they used this divine metaphor to show how Earth's very complex system has a life of its own in which we humans are but a small part.

Lovelock describes Gaia as "a physiological system because it appears to have the unconscious goal of regulating the climate and chemistry of a comfortable state for life... a whole system of animate and inanimate parts."[8] The Gaia model, therefore, "views the biosphere as an active, adaptive control system able to maintain the Earth in homeostasis."[9] Lovelock insists on using a relatable metaphor because he senses that unless people feel the Earth as a living system with a life of its own we will continue to respond to it carelessly and disrespectfully. The idea of Gaia is, therefore, a vital *aide pensée* and ethical prompt which allows us to realize that we have changed the Earth too much and so "she is about to move now... either by cancelling the changes or by cancelling us."[10]

Political theory is making an important Earth turn too. Political philosophers are rightly becoming more ecological and including the Earth and nature in what matters in politics.[11] They too are coming up with new political models which will inevitably influence humanitarian ethics and politics, and are already doing so. Several new terms have been coined to describe the new political age and the necessary politics we must shape as we respond to the Earth emergency. The best-known term is the *Anthropocene*, but the most accurate and useful term for shaping a new politics is the *Ecocene*.

Popularized in the 2000s, the term Anthropocene rightly recognizes that we are in a geological epoch in which humans have significantly changed the Earth in their own image and also have the power to destroy it to an extraordinary extent with bombs, climate change and environmental damage. Nevertheless, the political paradigm emerging from this geological insight still tends to centre human beings in political theory and views human life as dominant. Using a term first mooted by Rafi Youatt, from the New School in New York, the ecological philosopher Mihnea Tănăsescu argues powerfully that Ecocene politics is the best political model we have for our new age of Earth emergency. Tănăsescu prefers the idea of Ecocene to Anthropocene because "it has the benefit of putting ecology front and centre", explaining that:

> The irruption of ecological processes brings new kinds of actors into the polis... To focus only on the humans... misses the fundamental point that, in political terms, this new era is not about humans at all, but rather about how to accommodate, make peace with and negotiate with everything that is *not* human... and how to relate to it in regenerative ways.[12]

For me, the political model of the Ecocene is the right one. The challenge is how humanitarians, whose name and ideology are distinctly human-centred, can adapt ethically and operationally within a commitment to Ecocene politics and its ecological and multispecies values. Humanitarian ethics must also find their place in the emerging paradigm of ecopolitics and so work in pursuit of ecological justice and not just human justice.

As humanitarians imagine and work within the planetary scale of the Earth emergency, the scientific metaphors of boundaries and tipping points can help us to understand, reflect, advocate and act. So too can more poetic ideas of Gaia, or the Earth as our common home, and the new political paradigm of the Ecocene. The main thing is to raise humanitarian sights to a bigger picture and a wider ethic. In the long emergency of climate change, humanitarians need to conceive of their purpose in a new way and change their objectives and activities accordingly, so that they combine a new care for the Earth with their pre-existing care for humans.

In the Earth emergency, the humanitarian ethic cannot only be concerned with humane treatment between humans but must also involve a more humane treatment of nature and respect for the being of the Earth as well as for human being. It is not enough to work on human suffering in isolation. As Earthlings, as humans, we live as one part of a great array of life on Earth, much of which we love, most of which we need, and with all of which we should be especially careful. If we are to continue to live human lives, we must have much greater awareness and active moral concern for the Earth's system, which supports human life precisely because it supports all life.

Some humanitarians may feel that they do not need to make this ethical expansion. Caring for humans alone is more than enough for them and their agencies. For some, like emergency medics, search and rescue teams, and those visiting prisoners of war, this is largely true in their daily activities. But for a large proportion of humanitarians who are engaged in DRR, food security, public health, water supply, infrastructure support and nature-based programming, an Earth emergency perspective is inevitable and essential. How we integrate an environmental consciousness, compassion and practice into humanitarian action is, therefore, the main focus of chapters 4–10. In the next chapter, however, having begun to place great value on nature, we need to see how much it is changing and how disastrously it can hurt us and other life as it does so. We need to understand a little about climate-related hazards which will be the main cause of suffering, distress and death for humans in the Earth emergency.

# 3

# Earth's Elements as Human Hazards

Humans have always had an ambivalent relationship with fire. Just the right amount and it warms us and allows us to melt and modify all kinds of materials to our design and use. Too much fire and it burns us and everything around us. In Greek mythology, it was to avoid this risk of fire in human hands that Zeus banned humans from having fire, and why he was so furious with Prometheus, a lesser God, for stealing it and giving it to us. As punishment, Zeus pinned Prometheus to a rock and sent an eagle to repeatedly eat out his liver. But it was too late, and the energy given off by fire has been the basis of our emergence as a technologically sophisticated species. We have been burning away ever since—wood, peat, dung, coal, kerosene, animal fat, bees wax, whale blubber, oil, petrol and gas.

People have always known that smoke from fire was dirty and unhealthy in large doses. In the midst of the industrial revolution in Europe and America, it was very clear that the smoke from endless factories was polluting the air and creating thick smog in cities, which was making people ill. But it was a woman, Eunice Foote, who first noticed that something might happen to the planet if we kept burning away. In 1856, she realized that we might be making the world hotter as well as dirtier with all this steam and smoke.

Foote was a brilliant and enlightened woman from a progressive family of scientists and feminists in Connecticut and New York State in the USA. In her excellent history of the climate crisis, Alice Bell describes how Foote began to experiment with different gases to see how they reacted to heat from the sun. One of these gases was carbon dioxide ($CO_2$), and Foote was struck by its very different reaction to other gases. Her glass cylinder filled with $CO_2$ got much hotter than the others and took much longer to cool down. She wrote up a short two-page scientific paper of her results for presentation at the annual meeting of the American Association for the Advancement of Science. However, being a woman, Foote was not permitted to present the paper herself, so it was read out by a male friend. The paper was called "Circumstances Affecting the Heat of the Sun's Rays" and in it Foote prophetically observed of $CO_2$ that "an atmosphere of that gas would give to our Earth a high temperature."[1]

## Air

In the century before Foote's discovery, many scientists were identifying, isolating and naming the range of gases that we now know surround us. After Joseph Priestley in England identified oxygen in the 1770s, he and many other scientists were able to establish that what we call "air"—the thin atmosphere around the world in which we and other life can thrive—is made up of approximately 78% nitrogen, 21% oxygen and about 1% argon and a few other gases. After Foote, scientists like John Tyndall and Svante Arrhenius focused more on CO2 and its ability to warm the atmosphere. Gradually, it became clear that the heating effect of CO2 is not due to it making a hole in the atmosphere, like chlorofluorocarbons (CFCs) that destroy the ozone layer and let more rays of the sun inside the atmosphere of the Earth. Instead, CO2 works as an insulator and does not let heat out. The sun's heat hits the Earth in rays, but much of its heat then bounces back off into outer space as "blackbody radiation"—heat that is not absorbed by the Earth and emanates outwards again, leaving our atmosphere. As Foote noticed, the trouble is that CO2 blocks this escaping energy. This means that heat which should naturally leave the Earth stays within our atmosphere in a greenhouse effect trapped by what we now call "greenhouse gases".[2] This is why the Earth is getting hotter from humans burning fossil fuels.

CO2 from fossil fuels is not the only greenhouse gas. Methane and nitrous oxide also trap heat inside our atmosphere. Methane accounts for around 30% of warming, and most of it is produced by cows and other livestock in agriculture all around the world, especially in Brazil, India, China and the USA. Vast areas of forest are cut down to make space for cattle farming, most notoriously in the Amazon in Brazil. In the process, deforestation destroys carbon sinks (natural formations like the sea, forest and plants that absorb and store $CO_2$) and thus also increases $CO_2$. This is why eating less meat is climate action.

The difference between $CO_2$ and methane is their respective strength and longevity.[3] Methane is a much more powerful warming agent than $CO_2$: eighty-three times more powerful over twenty years and thirty times over a hundred years. However, while methane then breaks up, with 80% of it gone in twenty years, $CO_2$ stays in the atmosphere forever. This means that cutting methane brings about near-instant effects, while cutting $CO_2$ emissions slows warming but never cancels it.[4] Nitrous oxide is produced mainly from fertilizers and could be cut fast by a breakthrough in fertilizer manufacture.

In 1958, an American geochemist called Charles Keeling started measuring the amount of $CO_2$ in the air. Simon Clark's wonderful scientific history of Earth's atmosphere describes how Keeling began daily measurements of $CO_2$ in the air in remote national parks in California while he combined his love of hiking with his love of science.[5] As he walked in different isolated areas of the US, Keeling also took measurements in other parts of the country. In his early samples, he noted a steady baseline of 315–320 parts of

$CO_2$ per million of the air, as plants absorbed $CO_2$ and emitted oxygen. As Clark puts it: "Keeling was effectively measuring the planet breathing".

These measurements should have stayed predictable and stable, but the opposite happened and the ratio of $CO_2$ kept increasing. Over the next four decades, Keeling's observatory at Mauna Loa in Hawaii produced data in the shape of the famous "Keeling Curve" which consistently heads upwards. At time of writing, in January 2024, the ratio of $CO_2$ to air is 422.87ppm at Mauna Loa, which means that $CO_2$ is blocking far more heat bounce-back from Earth in 2024 than in 1959 and helps to explain our rise in average global temperatures. This expansion of $CO_2$ in the air combines both with rising methane and increasing water vapour to form the vicious circle of a warming world.

The expansion of $CO_2$ in the atmosphere coincides, of course, with the major global expansion in the industrialization and modernization of society enabled by energy from burning fossil fuels. If it started to speed up with coal, it accelerated much faster with oil. Known to people all over the world as an inflammable fuel for millennia, oil really took off in the mid-nineteenth century in Pennsylvania in the USA and Baku in Russia (in what today is Azerbaijan).

In Pennsylvania, on 27 August 1859, just two months after the Battle of Solferino in Italy where Henri Dunant had the idea for the Red Cross, Edwin Drake and Billy Smith struck oil at a depth of 21 metres. They did so with a small wooden derrick that was, ironically, based on a design for the gallows by the seventeenth century-English executioner, Thomas Derrick. His fatal name eerily describes pulley-based lifting and lowering devices throughout the oil industry. As one would expect in a contest between Russia and America, the Russians claimed to have struck oil in Baku ten years earlier in 1846.

Oil was a wondrous source of energy when first commercialized. In his monumental history of oil, Matthieu Auzanneau describes how, once found, it was a particularly easy fuel to source and move because it gushes of its own accord, flows in pipes and is poured easily into trucks and ships. Gas proved to be the same. Oil's energy density was also a game changer. In the late twentieth century, "oil delivered a 50:1 energy payback" so much greater than any other fuel.[6] As Auzanneau explains, this wonder fuel enabled a staggering development in lighting, heating and much more:

> This novel source of energy spawned entire new industries—notably the automotive, aviation and plastics industries, while revolutionising existing ones (agriculture, forestry, fisheries, shipping, manufacturing, lubricants, chemicals, paints, dyes, cosmetics, road paving and pharmaceuticals). It propelled humanity into an age of mobility and rising expectations.[7]

Oil made the modern world, finding its way into everything. The oil era coincided with "the world's population becoming the largest and most affluent in history".[8]

For many decades, oil and its energy were viewed predominantly as good. The devout pioneering entrepreneur of the American oil industry, John Rockefeller, felt himself on a mission, saying: "The whole process seems a miracle. What a blessing the oil has been to mankind."[9] Indeed, oil has brought public goods and personal benefits to billions of people across several generations. Yet it has also accelerated and intensified war, colonialism, pollution and the very global warming which is now threatening the stable hold of life on Earth.

As Zeus predicted, fossil-fuelled fire has indeed helped to burn and divide the world as well as to enrich it and render it more comfortable for humans. Our experience of coal, oil and gas has proved profoundly ambivalent. We now know—and since the Earth Summit in 1992 most governments have agreed that they know—that coal, oil and gas have poisoned and damaged the air and are causing global warming at a disastrous rate. Tragically, our main energy source has created a climate emergency in which Earth's elements are increasingly unleashed against us in a range of climate-related hazards.

These elemental hazards are driven by largescale climate oscillations and other atmospheric conditions, the "invisible giants" that shape Earth's weather.[10] Many of these are now increasingly "forced" into new patterns by anthropogenic climate change.[11] The conditions this creates can become disastrous for humans and ecosystems if they are dangerously exposed to them without proper preparation, adaptation and relief. In other words, if people and nature are critically vulnerable, and lack the capacity and resilience to endure such hazards, the changing climate will cause significant humanitarian and ecological disaster.

## Wind

Wind is air in movement in currents and waves. It is the atmosphere's way of transporting and distributing material around the world, especially heat, cold and vapour. The word weather itself comes from an older German word for wind, because wind makes weather.[12] For millennia, humans harnessed wind in sails of various kinds as a main source of our energy. Thankfully, we are beginning to do so once again at vast scale. There are regular winds and seasonal winds, like the Trade Winds, the Monsoon and the Sirocco, which are all different circulations of wind around the planet. There are also sudden winds that arise episodically, and there are circular "rotating winds" (known as cyclones in the Southern Hemisphere and hurricanes in the Northern Hemisphere) which track a certain path, often at huge speeds of up to 250 kmph, expanding and then breaking up as they go.

It is different formations of wind and pressure carrying heat, cold, rain, hail, snow and dust that make and move the storms, which as good weather and bad weather are productive or damaging to nature and humanity. High-speed winds are damaging and destructive in themselves, not only because of the other elements they carry with them and impose upon the Earth.

Wind and the contents of wind are vital to the Earth system, but they too are changing because of climate change. The warming of the oceans provides more energy to tropical cyclones, which are becoming stronger. Warmer water also increases evaporation and is producing more rain.[13] So we can expect winds to become more frequent, stronger and wetter. They already are; five recent storms have burst through the speed level for the maximum "category 5" wind speeds, so that scientists may soon need to agree new category 6 and 7 storms.[14]

## Water

The hydrosphere is a Greek term that scientists use to describe the total amount of water on the planet. This includes the oceans, seas, rain, rivers, lakes, aquifers, atmospheric vapour, snow and ice. Water is an essential element in the Earth system and many climate-related disasters involve either an excess of water in storms and floods, or an absence of water in droughts, heatwaves and wildfires.

The IPCC is predicting, with high confidence, that the Earth system will generate and experience increasing frequency and severity of flooding and inundations. This will be triggered by heavier rainfall, melting ice and rising sea levels. Flooding is the damaging flow of water, fast or slow, that drains away. Inundation describes water which stays on the ground to submerge a human and natural landscape for days, weeks or months.

In 2023, the extent of ice in the Antarctic Sea reached the lowest point since satellite monitoring began in 1979.[15] Counter-intuitively, perhaps, melting ice not only increases the risk of floods but also of heat. One way in which melting ice makes the Earth warmer is by changing the so-called Albedo effect of the planet. Named after the Latin word for white, this measures how much of the sun's heat is reflected back from white ice and snow around the world, and therefore not absorbed by the Earth. With less white snow and ice around the planet, the Earth absorbs more heat from the sun. This creates a vicious cycle in which ice that melts from warmer temperatures increases the amount of water on the planet. This means more water vapour, rain, surface water and rising sea levels, and bigger floods and storms. Less ice also means that global warming continues to increase as less of the sun's rays are reflected away from Earth. This is a classic cascading effect from a single change—melting ice—and a good example of a negative feedback loop that sees climate-related hazards increasing global warming by the damage they inflict.

Earth's warming is creating another water tragedy in the oceans of the world. Oceans cover seventy per cent of the Earth's surface and are extremely important for biodiversity and human food security. Oceans contain ninety-five per cent of the planet's ecosystems, soak up around thirty per cent of carbon dioxide in the atmosphere and produce half the oxygen we breathe. Over 3 billion people depend on ocean and coastal resources for their livelihoods. Like forests, oceans slowly absorb and sink huge amounts of $CO_2$ and seawater, while their plankton capture and store more $CO_2$ than anything else.

In 2023, ocean temperatures also reached their highest level in sixty-five years of monitoring,[16] with several "marine heatwaves" reported during the year. Rising sea temperatures cause several cascading effects. Firstly, warmer oceans absorb and store less $CO_2$ over time, leading to more global warming. Secondly, warmer oceans create more cyclones. Thirdly, warmer oceans damage biodiversity, including key foundations of wider ecosystems like coral reefs. Finally, of course, warmer oceans melt more ice sheets. Oceans and seas are not only getting warmer, they are also getting more acidic from the increased amounts of $CO_2$ they absorb. This acidification affects biodiversity and threatens or disrupts marine life in a variety of ways.

While we are experiencing more and more water in the world, the IPCC is also predicting with high confidence that some regions of the world will face extreme reductions in water for drinking and irrigation, largely because of heat. Increasing heat is not only problematic as a cause of more climate-related hazards in the hydrosphere. It is also a threat to life itself.

## Heat

2023 was the hottest year on record, breaking through the average increase of 1.5 degrees since pre-industrial records.[17] Humans and nature need heat, and most life thrives within relatively warm temperature bands. However, excessive heat can be fatal to the human body and to animal and plant life too. Human tolerance of heat is contextual to a certain extent. British men famously start sweltering, turning pink and taking their shirts off in summer temperatures of around 25 degrees Celsius; but this is a mild, comfortable heat for most people in Africa, Asia, the Mediterranean and the Americas. The UK Met Office recognizes a heatwave if there is a continuous series of hot days between 25 and 28 degrees Celsius, whereas most US states only recognize a heatwave at recurring temperatures of around 32 degrees Celsius. In India a heatwave is recognized at 40 degrees Celsius on the plains and 30 degrees in the hills.

Air temperature alone does not constitute the danger of extreme heat. The other crucial factor in a heatwave is humidity. If the weather is hot and humid, the body struggles more because less oxygen enters our system and less heat can leave it by sweating. Worse still, air pollutants accumulate more in hot and humid weather, meaning that there is not just hotter air and less air but also dirtier air that triggers respiratory disease. Heat stress and ultimately heat stroke are how medics describe the body suffering and key organs shutting down in extreme heat.[18]

Increased humidity poses rising risks for the 1.2 billion outside workers in the tropics.[19] But all humans struggle to thrive and survive in conditions of over 45 degrees Celsius, which is a temperature that is now more regularly being reached, especially in cities which can form "heat islands" and be hotter than their surrounding areas. These urban heat islands are caused by temperatures made even higher by the "impervious coverage" of urban buildings which prevents heat absorption and air flow, thus preventing organic cooling.[20]

The past nine years from 2015 to 2023 have been the hottest on record.[21] Within this aggregate rise in global temperature, spikes of extreme heat are becoming more frequent around the world and lasting longer, often as a simultaneous combination of drought and heatwave.[22] It is also clear that extreme heat is a direct killer in the climate emergency, perhaps the worst so far; heatwaves are often described as the "silent killer" of climate change by the International Federation of the Red Cross and others.[23] They are especially dangerous for outdoor workers and for elderly people and others with pre-existing health conditions, like heart disease and obesity, that make heat stress more risky. Emergency hydration and cooling programmes will be a key part of humanitarian programming from now on.

The combination of extreme heat and drought threatens global food security and increases water scarcity.[24] The world's wheat growing areas have seen heatwaves across a much wider surface area of production around the world in recent years, especially in the northern hemisphere. This area increase is matched by increases in the frequency and duration of heat events. Globally, between 1980 and 2020, extreme heat frequency was up by 28.2% and its duration up by 33.2%, meaning that droughts and heatwaves are already hotter and longer.[25]

*Fire*

Heat can damage nature because animals and plants struggle to cope in rising temperatures, but it also causes wildfires which destroy forests, drylands, ecosystems and human settlements. There are many different types of large fire. Some are started naturally by lightning or heat. Most are started accidentally or malevolently by humans, while others are part of careful human regulation of nature in farming or building. Wildfires are inevitably unplanned and burn according to the wind and dryness that keeps them going.

Fire is ecologically complex. A certain amount is good for nature. Too much is devastating. As with water, moderation is key and some wildfires or controlled fires can help to balance ecosystems. Fire ecologists agree that "fire is a powerful ecological and evolutionary force that regulates organismal traits, population sizes, community composition, carbon and nutrient cycling and ecosystem function."[26] Equally, however, they recognize that fire is increasingly a major challenge for human societies. Fire's impact on humans and nature includes direct damage and the death of people, trees, plants, animals, insects and ecosystems; habitat loss; health problems from smoke and pollution; emotional suffering; food insecurity; and wider economic losses.[27]

Fires are characterized by their scale, duration, intensity, behaviour and the severity of their ecological and human impact. All these features are getting worse so that OECD—not an organization that would previously have focused on fire—has declared that "extreme wildfires are a growing threat to humans, ecosystems and whole economies."[28]

OECD's report notes several grim trends. The duration of the wildfire season worldwide (the period when weather conditions are conducive to fire) increased by 27% between 1979 and 2019, especially in western North America, Western and Central Asia, large parts of Africa, the Southern Mediterranean, and Australia. The area affected by fire has also expanded. For example, in Russia it has grown fivefold between 2001 and 2021. The severity of the ecological damage caused by fires has also increased. In the

USA, severity has risen eightfold since 1985. Fires are also increasingly appearing where wildfire has previously been rare, such as in subtropical forests.

Scientists in Australia have concluded that the fires across parts of New South Wales and Victoria in 2019 to 2020 were "unprecedented" in their range, duration and severity. Some 5.8 million hectares were burnt in this period in a series of "mega fires", which burned 21.8% of Australia's temperate broadleaf forest. The reason for the scale, duration and severity of these fires was combined drought and heat. Fires rely on the extent of the "fuel load" available to them. This is the dry "litter layer" on the forest floor, which is made up of decaying branches, plants, bark and stem; this gets the fire going and then raises it into the trees. The drier the litter layer, the more combustible the forest. As Matthias Boer and Víctor Resco de Dios explain, "its dryness acts as the on/off switch for forest fire activity".[29] When the ground and the litter layer are moist, it is harder for fires to start, burn and spread; the gullies, swamps and slopes that dissect forests usually act as natural fire-breaks. In Australia in 2019, however, these were also hot and dry. In short:

> the litter moisture content across the eastern Australian temperate broadleaf forest biome was at record low levels, and the total surface area of forest exceeding critical flammability thresholds was larger and more prolonged than ever recorded in the past 30 years... These unprecedented fires may indicate that the more flammable

future projected under climate change has arrived earlier than expected.[30]

This trend for larger, longer and more severe wildfires has also been experienced in Indonesia, the USA and Greece in recent years. Like melting ice, wildfires are not just a problem of immediate ecological and human damage. They also add to general global warming, because burning emits large amounts of $CO_2$ that was sunk in forests and trees and releases it into the air again. The destruction of these precious carbon sinks reduces the future availability of nature-based capture and storage of $CO_2$, while more $CO_2$ enters and warms the atmosphere, cascading into further global heat entrapment, storms and droughts.

## Freeze

Many people around the world today are conscious that their weather is sometimes much colder as well as hotter. Cold waves and freezing spells are natural in many parts of the world in winter. The idea that global warming will bring on sudden freezes is not agreed by the IPCC, which predicts that, in decades to come, cold spells will be shorter and less intense. In the meantime, cold weather remains a climate-related hazard and humanitarian risk.[31]

In northern regions, cold waves and sudden freezes may be because of the "Cold Blob". This is a growing patch of cold water in the North Atlantic caused by melting ice. This colder water may be driving changes in the cold air that is firmly trapped above the Arctic by the Polar Vortex and the Jet Stream. A warmer Arctic and colder water may be causing the cold air pressure of the Polar Vortex to "meander", which makes the Jet Stream slip and cold air move suddenly southwards.[32] This in turn may be disrupting the oceanic Gulf Stream, which usually generates warmer air in Northern parts of the world to ensure mostly moderate winters.

Freezing weather can have serious humanitarian consequences, especially if people's homes and environments have been damaged by climate hazards or war, exposing them even further to the cold. This is why "winterization" programmes are such a feature of humanitarian aid.[33] The most comprehensive study of weather-related deaths suggests that cold weather is still a much bigger silent killer than heat.[34] This study looked across 750 locations in 43 countries and estimated that just over 5 million premature deaths occur globally each year from "non-optimal temperatures", with 8.52% being from cold weather and 0.91% from heat. This makes cold weather a significant hazard in the Earth emergency in the medium term.

### Soil

Finally, we must not forget the pedosphere. This is the thin two-metre film immediately under our feet, which we live on all the time and tend to take for granted. City dwellers tread only on an even thinner film of tarmac that we have spread over the pedosphere. Yet this thin film of Earth is not ocean, rock or metal; it is the mud on which, and from which, we humans and our neighbouring creatures live. It is the layer of the world that must be sustained if we are to eat and enjoy global food security as the climate changes.

The importance of the pedosphere for the generation, growth and survival of all life makes this delicate sphere of the Earth vital to humanity and nature. This soil on which we depend is, in turn, dependent on all the other elements to make life. It needs water and warmth, and occasionally fire. Human and animal activity can nurture it or damage it. Monocropping and artificial fertilizers can exhaust it. Relentlessly slashing and burning the vegetation on it can degrade it. Extreme heat can scorch soil of its fertility. Too much water or ice can submerge it, making it impossible for crops to thrive. The destruction of its trees can make precious soil slip and fall, or wash away, denuding the slopes and enriching the valleys. Under certain conditions, the pedosphere itself can become a hazard in mud slides, landslides, sandstorms and dust storms.

## Hazard risk

The climate emergency demands that humanitarians pay particular attention to air, wind, water, fire, freeze, heat and soil, and the worsening elemental hazards they are generating around the world. Nor do these hazards only strike one at a time. They often strike concurrently in pairs, trios or quartets to pose "compound, interconnecting, cascading and interacting risk" in the jargon of disaster management, and so create "multi-hazard risks". They cause "multi-hazard impacts" that demand "multi-hazard response".[35] Singularly and together, these elemental hazards have the potential to damage, degrade and destroy human life and the natural ecosystems in which all life survives and thrives.

Thanks to the long and determined work of many meteorologists, environmental scientists, disaster professionals and local communities working in the DRR field over the last seventy years, the humanitarian community knows a great deal about the character and impact of climate-related hazards. We also understand precisely how they cause great suffering and how to prevent and ease this suffering by reducing human and ecological vulnerability to such hazards. The next chapter looks at the ethics already embedded in the theory and practice of DRR. In DRR's concern for precaution, vulnerability and resilience, we already have three important policy principles on which to build a new humanitarian ethics for the Earth emergency.

# 4

## Precaution, Vulnerability and Resilience

The Yangzi floods of 1931 were probably the biggest single climate-related disaster of the twentieth century, killing about 2 million people. It serves as an example of the scale of climate-related mega-disasters and massive needs we may see if climate change tips Earth system boundaries into new levels of intense weather.

The winter in China in 1930 was extremely cold with unusually high levels of snow. In the spring of 1931, the snow melted to swell and flood the great Yangzi River that flows from west to east through large parts of China. The water table became dangerously high. Then, in July, seven huge storms burst across the valleys one after the other, and it rained and rained and rained.

Eighteen months of rain fell in one month. Inundation set in gradually. Flash floods struck with little warning. Dykes and embankments were breached. Slowly but surely, 180,000 square kilometres were inundated and a flood took shape that was 900 miles long by 200 miles wide.[1] One tenth of people in China—over 50 million people—were living in water. Many of these people were used to living *with* water and flooding, but this catastrophic flood overcame most of their aquatic coping systems, killing many and impoverishing most. It is estimated that 2 million people died alongside much higher numbers of horses, oxen, livestock, flora and insects.[2]

In his powerful account of this terrible flooding, environmental historian Chris Courtney describes why these floods were so disastrous. His book is a moving and analytical account of elemental hazards, human and nature's vulnerability and resilience, and of winners and losers among people and nature over many months of emergency. A similar scale of event today would involve even bigger human populations, a richer world, more technically complicated cities, and much higher buildings than in Wuhan and other cities in 1931. Today's human environment will have different strengths and weaknesses in the face of massive elemental hazards. For example, people may be able to stay drier and richer in China today, but getting water to people in thousands of high-rise buildings when supplies dwindle, electricity pumps fail or water freezes for days at a time is already a pressing new challenge in Chinese disasters.

Back in 1931, as Courtney explains, humans, flora and fauna struggled to survive in various ways.[3] For people, the sensory experience of weeks of wind, rain, wet and heat was physically and emotionally intense. The noise was constant in the "howling of the wind" and drumming of the rain. Repeatedly, as the flood spread, the "cacophony of rushing water and collapsing buildings" was heralded by human screams and succeeded by the dreadful sound of babies crying, or just a deathly silence. For weeks, bits of wreckage banged endlessly with "dull thuds" against walls and other wreckage. For millions of people in towns and cities, there was the ceaseless noise of lapping water where it had never been before in roads, houses, factories and shops. And suddenly, the clatter of gongs would announce fire spreading through an area because, bizarrely, the flood led to huge fires as collapsed electric cables and human carelessness ignited leaking chemicals and fuel amidst the wooden architecture. All around was the hum of mosquitoes, who surged and swarmed in such watery conditions, and the buzzing of flies who flourished on so many human and animal corpses.

There was also the smell. The "foul brew of flood water and sewage" filled people's nostrils for months, mixing with the constant smell of death. Looking at the flood became excruciating too, often simply numbing. The lack of dry ground made burial impossible. Corpses floated, swelled, burst and bumped around for months under people's eyes. The watery landscape meant that life turned upside down. People lived high up on roofs, walls and trees. Walking was limited to balancing precariously on beams, boards and branches, or wading waist high through water. Feet and legs were wet for months. People lost their normal sense of the earth beneath their feet and lived with the constant roll of water, becoming disorientated on firm ground. Everyone lived with "a palpable sense" of submergence and subversion. And, of course, it was dark. Electricity failed and candles ran out.

Experience of the flood varied for other species. Water birds, fish and frogs all thrived. Fish swam right into the city to enjoy vast new stretches of water and food, but were then more easily caught by hungry humans as large shoals swam through the narrow roads. Water birds could go wherever they pleased. Frogs went everywhere too as "the distinction between domestic and wild spaces" broke down completely. Mammals had a much harder time; millions of cattle, pigs and horses drowned or starved. Many microbes thrived. As mosquitoes surged, so did the malaria they carry, killing tens of thousands of people. Cholera rapidly "found a niche" and killed in tens of thousands too, alongside bigger killers like dysentery, typhoid and measles, which flourished especially as people fled to high ground and lived crowded together in large numbers. The heat and humidity were oppressive and exhausting. As so often happens in extreme disaster, "people did not have time to starve to death" and were killed much more quickly by disease.

People quickly became much poorer, deprived of assets, land, houses, livelihood and health. The pawn brokers and moneylenders were clear winners as people pawned tools, jewellery or whatever they had to get cash to survive. Huge increases in poverty and debt were the most obvious results of the floods for those who survived. Millions of people had to start from scratch and build their lives again when the waters finally receded, with many shouldering the plough themselves in the next planting season because so many oxen had drowned.

## De-risking climate hazards

From China in 1932, we can fast forward to Japan in 2015 to see how far governments have come in their efforts to understand risk and manage disasters without such terrible damage to life, livelihoods and assets. The test of the 2020s and 2030s will be to see if we can keep reducing the risks for billions of people, at the same time as hazards increase in incidence and intensity.

Disaster Risk Reduction is the field of humanitarian aid which sets out to make sure that hazards like the river floods and rain along the Yangzi in 1931 do not become catastrophic events. DRR professionals use an ethical logic that is now embedded in the principles and objectives of the International Framework on Disaster Risk Reduction agreed at Sendai on 18 March 2015,[4] and in many practical guide books on disaster management produced before and since.

DRR is not new. Humans have always reinforced their lives and homes against the elements. China has the longest continuous history of disaster governance that goes back thousands of years. Imperial and Communist rulers have always worked to reduce the effects of climate-related hazards, like drought and flood.[5] Since the 1970s, however, the science, ethics, practice and global recognition of DRR has developed at speed.[6] Grounded in the core concept of risk, today's DRR emphasizes an ethics of safety, precaution, community empowerment, and harmony and transformation in humanity's dangerous relationship with nature. To operationalize this ethical ambition, DRR focuses on a set of ideas and activities that include: vulnerability, exposure, prevention, mitigation, preparedness, evacuation, coping, capacity, community mobilization, nature-based solutions (NbS), adaptation and resilience.

DRR also uses models, and the best DRR model is still the "pressures and release" model from 1994.[7] This links root causes and proximate causes of disaster risk from climate change to political, social, environmental and economic pressures that construct people's vulnerability to natural hazards in different ways. For example, a poor family who live without decent work, land rights and assets in a low-quality house in a hazard-prone part of a coastal city is extremely vulnerable to storms and floods. If they are exposed to such hazards, it may be truly disastrous because their house may well be destroyed, their family members killed or injured, and any small assets lost or damaged. Conversely, the pressure of these same storms and floods is far reduced, perhaps only inconvenient, to a rich family with good jobs and significant assets who live in a strong well-made house set back from the coast in an area of the city with protective forest, good roads and public services. The model states that the core purpose of DRR is to release the social, political, economic and environmental pressures which make hazards so dangerous. This can be done by reducing the risk of people's exposure by improving their location and housing, and by boosting their capacity to cope with hazards by improving their livelihood and assets, and their political recognition and rights to state services.

In a sentence, the pressure and release model argues that *disaster risk* is created by a *hazard* whose impact is worsened by the extent of people's *exposure* and *vulnerability* to that hazard, which can be reduced by increasing their *capacity* to cope and resist the hazard risk. In other words, transforming people's vulnerability into more resilient capacity de-risks climate-related hazards. More simply still, the model is often expressed as an equation:

$$DR = H \times E \times V - C$$

Unpacking DRR ethics is important because they already contain vital ethical ideas that are foundational to how humanitarians should respond in the climate emergency. The DRR community has already done some, but not all, of the ethical groundwork for updating humanitarianism for the Earth emergency.

DRR ethics centres on the vulnerability of people and places, and is morally judgemental about the causes of vulnerability which it usually finds in unjust human relations and destructive human attitudes to nature. It also prizes precaution, empowerment, mitigation and adaptation. Ultimately, resilience is the goal of DRR ethics, a good enough end state if not a completely happy ending.

## Precautionary ethics

DRR is built on the moral importance of precaution, which is a forward-looking concern about dangers that lie in the future. Precaution gives moral value to the continuity of good things when we are not sure exactly how they will survive into the future. Precaution is an essential ethical technique when we are weighing up how best to preserve these good things in the face of an estimated but uncertain future, or in tempering the risks of an untried innovation.

Precaution recognizes the importance of foresight in ethics, which is "taking care of the future" and "safeguarding" valuable things for the future in the best ways possible within the limited knowledge of the present.[8] Taking precautions requires "pro-action" rather than reaction, the purpose of which is to *pre-empt* and *prevent* the worst that might happen to good things, or at least *mitigate* and *reduce* the estimated negative impact so it is not as bad as it could be.

The precautionary principle has a strong place in the ethical history of environmental ethics and was affirmed in

Principle 15 of the Rio Declaration back in 1992. Temperamentally and politically, precaution comes in weaker and stronger versions.[9] These reflect the two parts of the word itself. The weaker version tends to be overly cautious. It sticks closely to the status quo, acting only on high levels of future certainty and clear cost-benefit analysis. A stronger version focuses on the prefix of precaution (*pre* means "before" in Latin). Voiced most clearly by Henry Shue in climate ethics, this stronger proactive precautionary policy seeks deep change fast in the face of risks that may not be completely certain but which clearly risk "deadly delays" and "massive losses".[10] DRR aligns with the stronger version of precaution by urging significant action and investment as future-proofing on the basis of what we know already about climate change, its rising hazards and human risks.

By thinking ahead and taking a precautionary approach, DRR values two important features of the future that do not yet exist: future budgets and future life. There is a strong *value-for-money* argument in DRR ethics which follows the logic that investing now will probably save money in the long run of the climate emergency. DRR costs today may well be much cheaper than the ultimate losses and recovery costs of an unprepared society, especially as the intensity of hazards increase. This aspect of precaution shows humanitarians spending forwards. Studies already show that DRR investments are good value for money. IFRC estimates that every \$1 invested in their DRR programmes saves \$16 in lowered disaster response and recovery costs, and a global study led by the UN University shows that the benefits of precaution significantly outweigh the costs.[11]

DRR investments also target lifesaving over different timescales. Some prioritize immediate coping mechanisms for the current generation, such as: early warning, evacuation, preparedness or recovery grants to ensure that people will be safer when hazards hit. DRR ethics also goes beyond the present to think *intergenerationally* about lives not yet living. These activities are more structural, focusing on longer-term mitigation as well as adaptation that is intended to last for decades, including: improving infrastructure and shelter; rehabilitating ecosystems; livelihood diversification; and relocation. These are precautions taken now which will also help future generations.

## *The ethics of vulnerability*

In taking precautions, DRR's moral focus falls first on vulnerability. Human vulnerability, in particular, must be identified and reduced. This is clearly right. However, too often, vulnerability can become a blanket term that is routinely thrown over whole groups of people that may stigmatize them too. The designation of vulnerability can be damaging and disempowering, as well as prioritizing. The ethics of vulnerability therefore requires careful consideration, because vulnerability is something we all share as humans and will increasingly share in the climate emergency; this vulnerability also extends to all life, not just human life.

Catriona Mackenzie and others have usefully described three main types of vulnerability.[12] *Inherent vulnerability* is the bodily vulnerability we all share as humans who live, decay and die. In this vulnerability, we can hunger, thirst and be diseased, and be hurt physically, emotionally, socially and environmentally. Inherent vulnerability means we must depend on one another very significantly at certain points in our lives: for example, when we are very young, very ill and very old. *Situational vulnerability* is contextual, affected by our political, social, economic and environmental relations and how we can be hurt by discrimination, poverty, exploitation, or climate and environmental hazards. *Pathogenic vulnerability* is a more difficult term which describes a dangerous derivative of the other two, arising when solutions to our inherent or situational vulnerability put us at risk of new forms of suffering. The cure itself becomes a harm. A sad example here is a cognitively disabled person who is exposed to abuse by their carers. An example in climate emergency might be when people's relocation away from hazard risk turns out to be maladaptation, disastrous in different ways. In short, you are rendered more vulnerable by systems intended to protect you.

The social and political dimensions of situational vulnerability are concerned with people who face discrimination because of their ethnicity, gender, caste, class, age, ability, religion or views. Every society has people who are pushed to the margins of life because they differ in some way from the dominant group. In the margins, people usually also face economic vulnerability because they have the worst jobs and the least income and assets, as well as environmental vulnerability because they live in the worst places. They are settled beyond basic services, near dangerous water and de-forested slopes, and they may work long hours outside in heat or cold. This makes them highly exposed to climate-related hazards with few assets of their own to take precautions, and detached from government services which may support the resilience of more favoured communities.

A sense of autonomy is key in understanding and programming around people's vulnerability. Ethicists point out that vulnerability, and its sometimes inevitable dependency, are not the opposite of autonomy. If you look on me as vulnerable, it is important that you do not totalize this view and so completely "vulnerabilize" me. The humanitarian gaze always runs this moral risk of seeing only need and weakness and, when it finds some, can become intensely maternalistic or paternalistic in response.[13] Another dangerous tendency in vulnerability thinking is to see human vulnerability as purely negative, thus inviting social control of "vulnerable groups" by governments and aversion to them in supposedly non-vulnerable fellow citizens.[14] In truth, however, we all share a universal vulnerability as human beings. All our lives are contingent on the political society in which we live, our genes, disease, the actions of others, increasing climate-related risk, uncertainty and luck, and bounded only by the certainty of death. We all have vulnerabilities in one way or another, but we also have capacity and agency. An ethics of vulnerability, therefore, needs a nuanced view.

American philosopher Erinn Gilson rightly asserts how important it is that we do not stereotype vulnerability simply as people being susceptible, dependent, passive and weak in the face of harm and risk.[15] Highly vulnerable and dependent people have autonomy either directly as individuals who can easily make decisions, or in forms of "relational autonomy" with others who know, love or represent them, and so can share in their decision-making.[16] This recognition of strength and autonomy in vulnerability is important because nobody is invulnerable, especially in the climate emergency, but most of us also have autonomy and agency of some kind.

It is, therefore, more realistic and creative to see an openness to vulnerability as the beginning of wisdom and a potential source of strength. As Gilson reminds us, vulnerable comes from the Latin word for wound; vulnerability is the ability to be wounded rather than the wound itself. "This ability to be affected, however, leads not only to possible harm and loss but is also the basis for positive forms of connection and transformation."[17] Once we understand how we might be harmed, we begin to find power and common cause to prevent or reduce the risk, and can motivate others to cooperate with us as we do so. People who live consciously in their vulnerability live in a liminal space on the threshold of suffering that might and might not happen depending on what they decide to do. Self-awareness about our vulnerability gives us options.

In her study of ancient Greek heroes, Marina McCoy recognizes vulnerability as a virtue. Good personal and political action "depends on a certain understanding of human limit and vulnerability" which we all share.[18] In the climate emergency, an honest awareness and shared experience of vulnerability is a vital process in our efforts to avoid and reduce potential suffering from flood, fire, freeze and heat. Medical ethicist Joachim Boldt usefully terms this creative openness to our vulnerability as "valuable vulnerability".[19] It helps us to live wiser and more rounded lives in which we see ourselves and others more realistically, thus preparing and cooperating better to make the most of our strengths.

DRR ethics rightly insist that governments and humanitarians should recognize and empower the autonomy and strengths of people who are vulnerable rather than pathologize them. Working *with* rather than *on* vulnerable people is essential to discover the sources of vulnerability and prevent them from becoming disastrous, and to mobilize people's agency. This requires good community engagement, analysis, agreement and investment.

A positive understanding of vulnerability should also be extended beyond humans to the infrastructure and basic services which mitigate our vulnerability. It is not just people who are vulnerable to the elements in climate-related hazards, but also the infrastructure on which we depend. Extreme heat takes its toll on various kinds of non-adapted infrastructure like roads, runways, railways and dams which can be melted and cracked by heat, and on vehicles, power plants, machinery, food chains and medical supplies which overheat. Vital infrastructure and supply chains can also be blown apart by storms, washed away or inundated by floods. The World Bank recognizes climate-related hazards as a significant "drag" on the global economy because of losses in productivity, assets and infrastructure.[20] These may be direct damage to infrastructure and homes, or disruptions in individual livelihoods and across whole sectors like agriculture, travel and tourism which cease to function because of extreme weather.

In an Earth emergency we must also extend ideas of vulnerability beyond humans to nature, which has value of its own and provides our ecological infrastructure and services. Climate-related hazards can be disastrous for many species. One of the most detailed estimates of species suffering and loss comes from the recent Australian wildfires. One study estimated that 97,000 square kilometres of vegetation had been burned, composed of hundreds of floral species, and that this area was habitat to 832 species of vertebrate fauna. Of these, seventy types of species lost more than thirty per cent of their habitat.[21] A study by the World Wildlife Fund (WWF) attracted global attention when it estimated that "nearly 3 billion animals were in the path of the flames" suffering death, injury, hunger, displacement and long-term habitat loss.[22]

The humanitarian system routinely counts and analyses human vulnerability, suffering, death, damage and loss in disasters, but the vulnerability and loss of flora, fauna and ecosystems continues to go largely uncounted. The first attempt at a global overview of non-human damage, suffering and death from climate-related disasters shows these losses are a significant and integral part of every disaster.[23] Quite rightly, Yvonne Walz and other UN scientists leading this study have called for greater monitoring of ecological losses and recognition of ecosystems as "critical infrastructure" and a form of "basic services". It seems fairer to be even more explicit by ethically recognizing ecosystems as infrastructure and services for all species, and also for humans. This wider ethical view recognizes value and vulnerability in Earth's systems because of all life and not just human life, just as Chris Courtney did in his ecological history of the Yangzi floods at the beginning of this chapter. It makes moral sense to think of all life's vulnerability, risk and suffering in a disaster, not just our own.

DRR has started to make this ethical shift towards valuing nature, for itself and for us, by including nature-based solutions (NbSs) as part of risk reduction and coining two new acronyms to prove it: ecosystem-based disaster risk reduction (Eco-DRR) and eco-based adaptation (EbA).[24] These employ "natural infrastructure" and ecosystem services as important ways for people to find safety in resurgent nature, like restored wetlands and forests, and more resilient livelihoods from adapted agriculture. Working together, humans and nature can reduce the vulnerability of all life to climate-related hazards. As such, DRR is beginning to recognize the value of nature and make nature-based commitments in its ethics in a way that merges seamlessly with its original commitment to human life.

## Resilience as goal and virtue

Having identified nature's value, as well as people's vulnerabilities, autonomy and capacity, DRR works with humans and nature in their gradual empowerment and adaptation to reduce their vulnerabilities and risks so that they become more *resilient* to climate-related hazards. Resilience is the desired end-state of DRR and what is described as the new "resilience humanitarianism".[25]

Climate-resilient development (CRD) is also the new guiding star of the IPCC which sees it as an essential first step to sustainable development.

Resilience is an important and pervasive idea across all human societies and a feature of all life. Resilience makes sense intuitively to each one of us as we struggle to survive and live our lives. We also admire it in other life around us: the spider and her web; the migration of birds; and the hoarding and hibernation of squirrels. Resilience is found to be essential in all the natural sciences, in engineering, sport, and in social and political sciences.

As a moral goal, resilience does not suggest perfection; instead, the ability to endure, withstand and recover amidst imperfection. Resilience is not happiness or contentment because all is well and life is essentially good. Instead, resilience is a way of living, surviving and renewing when life is hard. In psychological work with suddenly disabled people, resilience is defined as "positive adaptation to a traumatic event" which somehow "transcends" the dominance of a disaster.[26]

In material science, resilience is strength, flexibility, staying power and durability. Material definitions of resilience have emphasized a bounce-back quality which sees a material put under stress but keeping, or soon returning to, its normal state without breaking. In engineering jargon, a material's resilience is thus measured on "its return rate to equilibrium upon a perturbation", or its recovery time.[27] We might imagine a tree bending in the wind or a dog's fur drying after a drenching. More than simply returning to normal, however, resilience also means adapting to new conditions. This would be more like a chameleon changing colour; sheep's wool producing lanolin to keep it dry in constant rain; penguins huddling together when it is cold; or someone rapidly learning a new language in a country where they recently arrived as a migrant.

Implicit in these examples of resilience are three other strengths: the ability to *absorb, deflect* or *resist* certain conditions. The bending tree absorbs the wind, the chameleon absorbs the colours of its environment, and the migrant absorbs a new language. The sheep's oily wool deflects the rain, and the dog's fur resists it and then shakes it off. Together these examples of resilience imply an ability to *tolerate* what might otherwise be intolerable. Resilience is about *strength, flexibility, toleration, adaptation* and *renewal* of different kinds.

In ethics, these strengths that make up resilience are recognized as personal virtues integral to good character and conduct in human life. They form key aspects of the practical wisdom, or prudence, that is essential to live justly and well, and are reflected in all the wisdom traditions of the world. They are: the insight and intelligence to understand; the courage to stand and act; the patience to absorb and tolerate; the flexibility to bend; the ingenuity to adapt; the ability to cooperate; and the fortitude, or determination, to endure when things are at their worst. As such, resilience describes core features of good character in all life as it struggles to survive and thrive in a changing world around it. In the climate emergency, these resilience virtues are essential for humans as we tolerate, survive, plan, cooperate, invent and reorganize society and our individual lives.

IFRC's definition of resilience is a useful one for humanitarians which captures these ideas:

> The ability of individuals, communities, organizations or countries exposed to disasters, crises and underlying vulnerabilities to anticipate, prepare for, reduce the impact of, cope with and recover from the effects of shocks and stresses without compromising their long-term prospects.[28]

Aiming for resilience in DRR and wider climate action insists, therefore, on a certain moral character of programming which embodies the various virtues of resilience.

More than this, resilience programming must maintain our integrity as human beings. We need to stay human and humane as we become resilient. Some sort of hyper-resilience or distorted resilience must not turn us into something else or something less than human. We must avoid becoming like the "preppers" in the USA who think only of themselves and are ready to live hidden away in a fully stocked bunker with rows of high-calibre rifles trained on potential intruders. Instead, resilience programming must always aim to preserve the uniqueness of human life and society so that we stay essentially social and the same while we bend, survive and adapt. Penguins are still penguins when they huddle, and chameleons are still chameleons when they change colour; they preserve their nature through resilience. This commitment to keep our humanity—and our respect for other human beings—must be central to resilience programming, so that resilient humans remain truly human and humane while they struggle to survive.

Such *values resilience* is an essential measure of success in resilience and adaptation. We should not change the benevolent nature of human beings or reduce human society to a lesser moral state as we struggle and adapt. Programming which simply supports the resilience of one group at the expense of another would be unkind and unfair. Resilience building that imposes inhumane resilience by exaggerating risk, deliberately stoking fear and inhibiting freedom would be unethical.[29] An excessively controlling resilience regime might crush human relationships, exploit nature, and inhibit the freedom and creativity we need to adapt justly and well. Resilience programming has a responsibility to cultivate the virtues of resilience while safeguarding human integrity and the ability to live a distinctly human life in harmony with nature.

This chapter has shown how the ethics and ambitions of DRR, and its key concepts of precaution, vulnerability and resilience, are hugely valuable moral guides to humanitarian action in the Earth emergency. They already represent a significant evolution in humanitarian insight. But we also need to go further today. The Earth emergency demands that we also update the first principles of humanitarian ethics. The next chapter begins this process with a reformulation of the principle of humanity itself.

# 5

# Humanity and Nature

The Earth emergency requires humanitarians to look beyond human life alone and recognize the unity and value of all life. This change of view means a shift in humanitarian ethics, which must start with a moral deepening of humanitarianism's most fundamental principle, the principle of humanity. This is currently framed just as it was in 1965:

> to prevent and alleviate human suffering wherever it may be found... and to protect life and health, and ensure respect for the human being.[1]

In the Earth emergency, a humanitarian commitment that is confined solely to the human species is ethically incomplete and ultimately self-defeating.

In this chapter, I argue that a proper concept of humanity in the Earth emergency needs to involve a profound ecological awareness and environmental purpose that is born out of a closer human identification with the Earth, and compassion for all life. An updated humanitarian morality must see humans as an integral part of nature, dependent on nature and bound up in shared suffering and survival with nature. A mission to protect only humanity in an Earth emergency is flawed because it is isolationist, unrealistic and ultimately doomed. Instead, humans must be protected in nature and with nature, and humans are simultaneously responsible for human life and all life wherever they have power and capacity to act.

The most ambitious attempt yet to update humanitarian ethics in this way that I have found is, surprisingly perhaps, in a master's thesis, by Reema Chopra from the IFRC.[2] Chopra's paper is a comprehensive "holistic vision of humanity" which "reinterprets, expands and re-envisions 'humanity' through deeper environmental consciousness… extending without discrimination compassion and efforts to prevent and alleviate animal suffering… and ensure respect for the diversity and dignity of the Earth."[3] Chopra and I share the same commitment to compassion for the Earth, and I use the same idea of "deepening humanity" but in a slightly different way. Chopra's work is pioneering, and her ethical sweep extends from the inner consciousness of every human person to an active concern for all life. It is an impressive call for a humanitarian paradigm shift.

Making this moral shift means reassessing our very sense of humanity itself to better recognize its embedded relationship with nature. In the new Climate and

Environment Charter, humanitarians rightly affirm that they need to pursue a more earthy and environmental approach to their work. I hope this chapter helps to explain the ethical reasons for this by describing a more symbiotic and mutualist understanding of the relationship between humanity and nature, and then recommending three possible revisions to the principle of humanity in the next chapter which would bring it up to date.

The ecological shift in the humanitarian mindset starts with a deeper appreciation of life itself and what it means to be human amidst other life.

## Life

When increasing heat poses such a pressing threat to life on Earth today, it is strange to think that life may first have started in the 90 degree Celsius heat of alkaline hydrothermal vents deep under the sea about 4 billion years ago. Life stayed at the bacterial level for 2 billion years until around 1.5 to 2 billion years ago when a single complex cell seems to have emerged which somehow made a complete break out from bacterial form. Summarizing what we know about the origins of life, the evolutionary biologist, Nick Lane, explains:

> All complex life on Earth shares a common ancestor, a cell that arose from simple bacterial progenitors on just one occasion in 4 billion years... This common ancestor was

already a very complex cell. It had more or less the same sophistication as one of your cells, and it passed this great complexity on not just to you and me, but to all its descendants from trees to bees… This ancestor was recognizably a "modern" cell, with an exquisite internal structure and unprecedented molecular dynamism, all driven by sophisticated nanomechanisms encoded by thousands of new genes that are largely unknown in bacteria.[4]

Scientists do not understand for sure how this breakout happened, and have found no missing links to trace its gradual emergence. However, along with our common cellular ancestor, the branches of the tree of life—of which we humans are a part with plants, animals, fungi, seaweed and amoeba—began to evolve and grow. Those simpler forms of bacterial and archaeal life stuck to their two different routes and continue to exist, helping and threatening us in various ways.

Today, as far as we know, all forms of terrestrial life inhabit a thin zone of life that ranges from three kilometres below the ocean floor upwards into twenty-three kilometres of the Earth's lower atmosphere. All this life takes different shape in 10–30 million species.

*Being*

So, we are not the only ones alive. But we are the only ones alive as humans. Every living creature experiences their form of life in a particular way. This creates a certain distinction between life and being. At one level, life is as simple as NASA's universal concept which defines it as "a self-sustaining chemical system capable of Darwinian evolution." Being, however, is the feeling, agency, experience, and peculiar limits of living as a unique form of life. In Greek philosophical terminology, living is our *biology* and being is our *ontology*. They are, of course, integrally related. Biological inevitability means that my being human is different to an oak being a tree, or an eagle being a bird.

As different creatures, our senses, perceptions and meanings of the world are different and unique. The way of being-in-the-world is particular to each species. Jakob von Uexkull, a pioneering twentieth-century biologist, used the term *umweld* to describe the distinct lived environment and worldview of each different species, and he famously tried to sketch the worldviews of a tick, a sea urchin, a dog and many more.[5] No species is the same as another or experiences the world like another. We humans can overlap with animals, in particular, through our five senses and some of our reasoning about the world. But this overlap is very small. American zoologist Katy Payne, and others after her, have discovered that we miss so much animal communication that passes us by as infrasound. Whales, elephants and many other species talk to each other in sounds below or above our hearing range, and in signs and languages we do not understand. For example, an elephant sound, which we cannot hear, can fill 300 square kilometres of African savanna and elephants can communicate easily with one another across a range of 16 kilometres, hearing through their feet or through their ears.[6] The world is full of animals communicating with each other, most of which we cannot hear, see, smell or feel.

As humans, we may empathize with what it feels like to be a salmon leaping and swimming upstream, but we can only ever imagine it as we might feel it in our own bodies, and not as the salmon really feels it. We can also feel the distinct being that trees, plants, flowers, birds give off, excited that it somehow shares and celebrates our deep sense of being too. Yet we still do not really know what it is really like to be a flower or a bird, and cannot tune into how they communicate with one another. The Holy Quran beautifully affirms this reality of our different ways of being on Earth: "all the creatures that crawl on the Earth and those that fly with their wings are communities like yourselves."[7]

Like many of us, I have had moments of epiphany in nature. Sometimes these involve feeling at one with the world in some eternal sense. At other times, they involve sensing the extreme difference of being in an animal, plant, mountain or ocean that is alive in such a different way to me. This is exciting for how *unhuman* it is. On a walking safari on a small island at Naivasha in Kenya, I spent time standing still beside groups of zebra and giraffe, hearing their quiet chomping, and occasionally held in their evaluating gaze. I felt their being so palpably different to my own. Zebra being and giraffe being rippled through time and space so strangely: humming at a different tone that was steadier, more collective and more knowing of its world. Their different pitch of being has stayed with me as insight and exhilaration. It tells me that the human world is not the only society on Earth, or the only way of being, and that nature and animals *are* just as we *are* too, but in a different key.

Robin Wall Kimmerer, an American botanist, calls this "the wordless being of others" in which we are never alone.[8] She talks of the "animacy" of nature to describe the aliveness and universal being we all share. Kimmerer describes how many human groups recognize this animacy in their languages by giving animals, mountains, forests, rivers, lakes and seas a form of personhood as he or she or they. This linguistic habit recognizes nature and other life as "someone" and not "something", and she quotes the Catholic theologian, Thomas Berry, who urges: "we must say of the universe that it is a communion of subjects, not a collection of objects."[9]

Humanity's shared being and shared space with nature has consistently been expressed in sacred and poetic terms.[10] All religions and most wisdom traditions tell of how all creatures and all natural forces share in an essential being. Ancient Chinese wisdom recognized this as the Tao or Qi, which runs the world and runs throughout the world in animals, rivers, rocks, flowers, winds and human lives. Many traditions talk maternally of this shared being as the mother of the world, and some speak of it as the father or spirit of the world. Thomas Aquinas talks simply of shared being as God being in all things.[11] The Upanishads describe a universal essence or spirit, Brahman, that "enfolds the whole universe, in silence is loving to all... and is the Spirit in my heart."[12]

The Ethiopian environmental philosopher, Workineh Kelbessa, describes how various African philosophies and theologies express this sense of shared being.[13] The totemism in many parts of Africa believes that humans, animals and plants share an important spiritual respect and connection with one other, so that many groups take the name of animals and plants as their clan names. Many African philosophies also express the unity of life and being in important doctrines about the Earth. For example, the Oromo in Ethiopia recognize the unity of *Waaqa*—Earth and human beings. They insist on *Saffuu*, which holds that all things have a place of their own and must be held in balance, without violation.

In modern environmental politics, these sacred understandings have been brought powerfully into the political arena. The anti-capitalist Indian scholar activist, Vandana Shiva, has developed Hindu thinking to argue for "Earth democracy" in which there is justice for all species and a new conception that all life and nature lives as one "Earth family".[14] Under the leadership of Bolivian President

Evo Morales, the Universal Declaration of the Rights of Mother Earth was agreed by certain states, indigenous peoples and civil society in Cochabamba in 2010. It is a powerful affirmation of the political vision of the Earth community which declares that "each being has the right to a place and to play its role in Mother Earth for her harmonious functioning."[15]

The sheer beauty of nature and other species is universally experienced by all human civilizations. Humanity is dazzled and delighted by the colours, shapes, and scents of the non-human world. All human cultures imitate this beauty in art, dress and design. This beauty uplifts us and infuses us with joy. As Carl Jung and others have revealed, nature's reflection has settled as common archetypes and symbols deep within our human consciousness with which we feel the world in nature's image and express ourselves in natural metaphors like blue-sky thinking, green energy and tempestuous times. In this way, nature provides many of the building blocks of our imagination. The beauty of nature and its imaginative presence in our minds shows us how valuable it is as something to love and not to lose. The beauty of the Earth, like the face of a friend, is precious to humanity and a vital part of us.

Sadly, there is an equally long history of denial and disregard for this sense of beauty and shared being with nature. Human greed and excessive predation have ravaged the natural world around us. Many modern ideologies of human supremacism, like extreme capitalism, fascism and communism, have politically separated human life from nature so as to exploit it and degrade it. This *reification of nature* in human ideology dismisses non-human lives and being as merely *things* that have much less value than human lives. Instead, non-human life and matter hold only *instrumental* value as useful or not to meeting human needs and desires. Such supremacist thinking leads to the *bifurcation of nature* in many societies in which we split ourselves off from nature. In the process, we discount the being and society of other life and matter to give primacy to human importance and our inevitably partial human view of the Earth.[16]

## Togetherness

Humanity is nothing without other life. Not only do we humans share life, and an experience of being, with a range of species and natural matter, but we are all intricately linked in the processes which make and sustain all life. Pope Francis notes how we simultaneously hear "the cry of the Earth and the cry of the poor" because "the human environment and natural environment deteriorate together". This means that "a true ecological approach always becomes a social approach" and so vice versa, a truly humanitarian approach becomes an ecological approach.[17]

Modern ecology has given new insight into humanity's inter-dependent and embedded relationship with nature. In 1935, Arthur Tansley, a botanist at Oxford University, coined the new term *ecosystem*. With many others, like Charles Elton working in zoology, ecologists soon began to show how all life forms are interconnected and interdependent in intricate ways. This systems perspective allowed the new field of ecology to identify scientific concepts like food chains, energy flows, regulatory pathways, ecological pyramids, evolutionary niches and invasive species.[18] The way each part of nature lives as different spaces, energy and creatures determines or denies the existence of other species.

The idea of ecosystems is not really a discovery. Rather, it is an analytical model that has enabled modern science to rediscover the integral relationship between humanity and nature that was long understood by ancient and traditional peoples. This unity and interaction of all life was long valued by many societies in the moral principle of *harmony* between humanity and nature. Harmony is not simplistically understood as existence without contest or violence, but as a proper *balance* in which no single species and no natural object, like a river or a wind, should take or destroy too much. In the philosophy of Shona people in southern Africa, the principle of *Ukama* represents this moral and environmental insight. *Ukama* means relatedness. It stresses the vital embedded and intimately connected relationship between nature and humans. Without balance and harmony in this relationship, neither can survive or justly thrive, and their environment will fail.[19]

Anna Tsing, an American multispecies anthropologist, has worked on the interconnectedness of life and come to think differently about our modern human ideas of species and individual ontology.[20] Tsing works a lot on fungi. From mushrooms of many kinds, she has come to understand that fungi are always living with, and adapting to, other species as they grow and move around the world. Her insight into the close relationships of different species and landscapes, which sees them flourishing together, leads her to conclude that there is no such thing as a single ontology, a completely individual way of living and being in the world. Every species *is* by virtue of how other species *are*, and how we opt and evolve to be together, or not. Therefore, Tsing suggests that we all experience a *relational ontology*—a way of being that is as much about togetherness as separateness. In other words, the being we are is co-created, so that life is always *lived with* and *lived among* and *lived for* other life around us.

This is true for humans too. In our humanity we always live hybrid lives and inhabit relational ontologies with other forms of life, which makes us *not-just-human* and shapes our humanity so that we live in a web of shared being that is *more-than-human* too. For example, Tsing shows how humans domesticated wheat but how wheat also domesticated humans in the process by enabling us to settle.

The consequence of our close relationship with wheat, and our living as wheat-humans, was to co-create new forms of landscape, habitation and governance, with wheat surpluses even enabling us to invent taxation and govern differently. Thus emerged a distinct version of humanity in which humans and wheat co-created a new way of being human. Being rice-humans has done the same across large parts of Asia. Being dog-humans has enabled us to hunt. Being horse-humans, wind-humans and sea-humans has accelerated trade and war. As we saw above, being oil-humans has seen us enriching and industrializing people of the Earth, and also damaging the Earth and dividing people.

Tsing's insight into relational ontologies—or mixed being—shows how humanity is never simply human in this world. Biologists also show us this truth in our own bodies, in which we are never just singular human individuals. Our gut alone is home to millions of other lives and so we, like all other species, tend realistically to live (in a rather clunky new term) as a *holobiont*—a biological assemblage of our life as host to many other lives.[21] Our human lives and our human being are constantly *symbiotic* and involved in *mutualism* of various kinds with other life on Earth.

There is a lovely Muslim hadith which poetically conveys this truth about humanity's integral embeddedness with nature, showing how our humanity is always lived, and made richer, in relationships with different life around us. It is said that the Prophet (PBUH) used to give his Friday sermons standing beside a date palm tree, often sitting on an old palm stump. As he talked, the tree leaned in to shade him and to enjoy the beauty of his words. Soon, the people made a formal pulpit for the Prophet and he left the tree and stump to speak from the pulpit. But the tree became very sad, missing the Prophet and his words, and the stump began "crying like a camel", so the Prophet came back, rubbing his hand over the tree stump to soothe its crying, and preaching close to the tree once more.[22]

## Reimagining humanity

What does all this ecological philosophy, anthropology and theology mean for humanitarian ideology and the current principle of humanity that governs humanitarian aid as its primary moral goal? Today's recovered sense of ecological awareness, so accelerated by environmental movements, Earth science and the climate emergency, gives us an important opportunity to deepen our understanding of humanity and reimagine our first principle of intention and action.

The Earth emergency offers an inspiring moment of ethical renewal. Humanitarians understand our togetherness as a species in the sense of our common humanity. We now need to go further and recognize the ecological and moral truth of our common life with nature. If we do not, we risk misunderstanding what humanity is. The next chapter explores how we can revise and renew the first and most fundamental principle of humanitarianism so that it is fit for purpose in the Earth emergency and represents a truer version of ourselves.

# 6

# Deepening Humanity

The intellectual historian, Bruce Mazlish, has described how centuries of discovering each other around the world and map-making a common globe to represent the Earth has enabled humans to construct and achieve "the idea of humanity". This has given us a global sense of the first-person plural: a new "we". Citing the sociologist, Jose Casanova, Mazlish notes that the festivities to mark the year 2000 were probably the first time that the whole of humanity had celebrated the same event together.[1] This recognition of a common humanity has crystallized a new level of universal human consciousness in our lifetime (represented in the idea and institutions of the United Nations) even if this consciousness is fragile and conflicted.

The Earth emergency rightly calls us to deepen and expand our understanding of humanity still further to create a richer "we" that includes all life and the rock, water, air and soil of nature on which we all depend. In the words of Thomas Berry, humanitarians must deepen our sense of self, compassion and responsibility by "rejoining the Earth community" and recovering our "human intimacy with the Earth".[2]

Clear recognition of the human place in nature complements the shared feeling we have found in Earth's human community. It also means adjusting the identity and purpose of humanity to recognize nature-based commitments as well as human-based commitments. China's President Xi urges that "humanity must protect, respect and stay in harmony with nature."[3] In an Earth emergency, we can only fulfil our human commitments if we recognize our Earthly identity and meet our commitments to nature.

This realization that humanity is part of nature and has responsibility for nature is honoured in one of the oldest stories we have in the world—the ancient Mesopotamian story of the flood. This disaster myth is found widely across the ancient world and comes down to us in our oldest written human text, the *Epic of Gilgamesh*, and later in the *Book of Genesis* in the Jewish Torah. In both accounts, the chosen human survivors of the great flood, Uta-Naphisti in Gilgamesh and Noah in Genesis, along with their nameless wives, are tasked by God to save themselves *and* all living things so that Creation can be renewed.[4] Like Noah and Uta-Naphisti, we need to take all life into the humanitarian ark or the Earth will be unliveable.

## *Renewing humanity*

Humanity's renewed sense of the importance of nature brought about by climate change and the Earth emergency means we must revise and renew the principle of humanity.

In our current understanding of humanitarian ethics, humanity means three different but related things: a species; an identity and a virtue, which we might call humanity meanings 1, 2 and 3.[5] Humanity (1) refers to our species, *homo sapiens*, and describes our particular form of biological life and way of being in the world. Humanity (2) refers to our common sense of human identity as a global way of life that covers the Earth and can recognize and communicate with each other. Humanity (3) refers to being humane, the virtue of human compassion, kindness and collaboration, which humans exercise when they "show humanity" to one another.

To revise the principle of humanity for humanitarian action in the Earth emergency, we need to adjust our understanding of each one of these three meanings of humanity.

### Humanity as species

Adjusting the first meaning of humanity—species humanity—involves affirming what is common and unique about humans as a form of life on Earth. As a species, we are an animal and share much of our physical, emotional and rational make-up with other animals. Just as it is important not to reify nature, so it is equally important not to deify humanity. Animals, and in particular mammals, are the wider group to which we belong on Earth and we have important things in common with them. British philosopher Mary Midgley pointed out in her seminal work on ethics and animals that: "human life really does have an animal basis—an emotional structure on which we build what is distinctly human" and this means that "quite complex aspects of things like loneliness and play and maternal affection, ambition and rivalry and fear, turn out to be shared by other social creatures."[6] We mix with animals and are similar in important ways to many animal species in how we live (by eating) and how we interact (socially). We are of the Earth like other animals and we need the Earth like they do. We can know and love other animals, and be known and loved by them.

We are also different to other animals for two main reasons. First, because we have developed our reason and technology to such an extraordinary degree. Secondly, because this has given us the capacity to shape what happens on the Earth at a local and a global level. This begs the question of our status and role in the Earth community. Our species is certainly distinct among animals in its coverage and power on Earth, but what is the ethical nature of this distinction? Are we biologically exceptional somehow? Certainly not on an individual anatomical level, because each one of us dies and decays like every other animal and species of life. Are we exceptionally important hierarchically so that we can rule the Earth, taking from it whatever we like and using it in any way we decide is best? Then, perhaps leaving it one day to fly off to another planet that we have prepared as habitable? In other words, is our exceptionalism one of intrinsic supremacy over all other life?

The question of human exceptionalism is essential to resolve in a new humanitarian ethics in order to agree the significance, rights and duties of humanity in the Earth emergency.

It makes most sense to resolve our species status in a middle way between two extreme positions. In contrast to the idea that we are simply a part of nature, there exists a long-held view that humans are in fact a species set apart, rightly dominating the Earth as a semi-divine super-species. This extreme version of human exceptionalism is morally wrong because it conflicts with how most of us feel in the world around us, especially when we stand in awe of its beauty and greatness, when we are rendered terminally ill by a virus or a cancer, or left destitute when our house is destroyed by a storm.

Human supremacism is equally unlikely because our so-called rule has clearly not been in the interests of the Earth as our kingdom. As things stand, our exceptionalism looks untrue and self-defeating. Added to this, many humans have also suffered terribly from this dominant mindset which, politically, is often deployed in the interests of a few and has regularly created human injustice and atrocity. Finally, of course, the obviously dire consequences of our (mis)rule seems to be destroying the world as we need it, and so is proving to be suicidal. The idea of humans *dominating* the Earth seems to be going disastrously for us and the Earth.

The other extreme reading of human exceptionalism is to reject it completely. Many deep ecologists, for example, insist that there is nothing exceptional about humanity as a form of life except that we are exceptionally dangerous to the Earth. They think that humanity's value is equal with animals, insects and microbes. Some go further to think that the best thing for the wider Earth community might be for us to be wiped out, or dramatically reduced, as a pest and invasive species of some kind. They expect Gaia to reach the same conclusion soon.

A more reasonable view of human exceptionalism is best built around particular human responsibility rather than exceptional human status. Our uniqueness in the Earth community is found in our reason and capacity to understand and manipulate some of the world around us, and thus our responsibility to care for it. For a long time, some in the Christian tradition have misunderstood Judeo-Christian theology to believe incorrectly that humans are the pinnacle of God's creation—its glorious high point. In fact, this is a narcissistic reading of the texts. Metaphorically, we may be "made in the image and likeness" of God and "a little lower than the angels", but we are not gods.[7] Like us, all nature expresses the glory of God, or less religiously, the creative spirit of the universe. Humans are not the best thing about the Earth but one of its most unique creatures.

Ecological theologian Celia Deane-Drummond explains how in Jewish scripture the culmination of creation is not humanity but the Sabbath. In the Torah, the Sabbath is the seventh day of creation when *everything* is finished: the heavens, the Earth and all living beings are formed out of chaos. At this point of completion, God rests and rejoices in everything created, and enjoys it all as good. "It is the Sabbath rather than humanity which is the crown of creation" and humanity's unique purpose is found within and with nature to care for the Earth.[8] When Jewish families rest and eat together on the Sabbath as the seventh day of the week, they are celebrating the creation of the universe, the Earth and all that is in it, relaxing from their particular responsibilities within it rather than over it.

An updated reading of humanitarian ethics must situate humanity firmly within nature, and dependent on it in various forms of mutualism. It must be more honest about human exceptionalism as uniqueness rather than supremacism. It must also recognize humanity's particular duty of care towards all life on Earth because humans are a unique member of the Earth community that has achieved global reasoning and widespread Earth responsibility.

Human nature and human purpose clearly reach out beyond human life alone. Being human means being part of the Earth, as well as individual and transcendent, and human morality is not only a matter of caring for other humans in a single species ethics. Humanity involves a duty, a desire and a need to care for all life.

## Humanity as identity

This fuller definition of the human species within the Earth community, and with responsibilities for it, immediately changes our sense of identity as global humanity. Our identity stays uniquely human, and we continue to *identify as* humanity. But our place in nature, our relational ontology with so many other forms of life and matter, and our particular responsibility to care for the Earth as much as we can, deepens what and whom we *identify with*. In other words, a richer sense of humanity extends our familial and ethical sense of "we" and widens the moral community of which we feel a part.

In the words of the French sociologist, Bruno Latour, this deeper sense of identity with nature brings us "down to earth".[9] It makes us properly aware of the other forms of life and the terrestrial landscape on which, and with whom, we live and thrive. It is with them and with the Earth that we must find equal and common interests. In doing so, Mihnea Tănăsescu reminds us that "mutualism is a feature of the living" and of the "myriad ways in which life is only possible because of cooperation".[10] In arguing for a new sense of mutualism in our sense of humanity, Tănăsescu quotes the biologist Lynn Margulis' revealing phrase that "independence is a political, not a scientific term" and it will not do humanity any good to seek environmental independence.[11]

We can never operate only as humanity because we belong to something bigger—the Earth community as well as the human community. As humanity, we always live as a "mixed community" of species, interacting socially, economically and emotionally with them in various ways.[12] We depend on one another and so are effectively one tribe. In short, we share a common identity with nature as well. This gives us obligations to humans because of our common humanity, and to all life because of our common life and common home.

This growing sense of the communal significance of other life around us is manifest in humanity's increasing recognition of nature in law and politics today.[13] As we value nature more deeply as part of our moral and political community, so we share our legal identity with it. Various national constitutions and legal rulings are giving legal personality to other life and matter, and establishing the "rights of nature" (RoN) alongside human rights as part of what humanity should care about. This legalizing is humanity's attempt to represent nature in our politics, as part of what matters to us.

In the 1960s, Hanna Pitkin famously defined representation as "making present that which is absent". This can be done by delegating, trusteeship or presenting claims on behalf of others. The rights of nature often build explicitly on indigenous people's worldviews and traditions. In New Zealand, the Whanganui River now has legal personality represented by a board which is the "human face of the river." These modern legal developments are themselves new versions of what Mary Midgley calls the many "tacit agreements" we have with nature and animals, which are genuinely made and deeply held even when we cannot talk to one another.[14] They form a natural part of the extensive non-verbal communication we enjoy with animals, plants and humans who cannot talk, or do so in a different language that we cannot understand.

The Ecuadorian constitution is perhaps the most progressive in its recognition of the rights of nature. It builds on the traditional Ecuadorian principle of *buen vivir*, which is informed by the earlier Quechua concept of *sumak kawsay* and means "living well" in dignity and fulfilment.[15] The constitution affirms "the right of nature to have its existence respected holistically, and to the maintenance and regeneration of its vital cycles, structure, functions, and evolutionary processes."[16] In various judgements, the Ecuadorian courts have gone on to declare that mangroves and rivers are "the subject of rights" and ordered their protection. These judgements are creating important precedents about the intersection of the rights of nature and human rights. Here, the tension will be whether the rights of nature are used mainly for guaranteeing human rights, or whether nature's rights will ever be firmly upheld against human rights. New political actors—in this case nature—inevitably add to political contest in some form. However, nature is at least now formally inhabited and represented in the human world, just as humans have always inhabited and imposed themselves on nature.

## Humanity as virtue

The third meaning of humanity refers to the virtues of compassion and kindness that humans show when they are being humane. Here, it is obvious to point out that humans are not always humane and can be profoundly and extensively inhumane, which is why the humanitarian project needs always to be consciously adopted in norms, laws and institutions. It is also important to note that the human species in no way has a monopoly on the virtues of kindness and compassion, which we so self-referentially define in our own image as the virtue of humanity. Many species are humane to one another in the way we define it. Some species are actively kind towards us. They initiate kindness and do not simply reciprocate our kindness to them.

The adjustments to humanity as species and identity demand a third adjustment to humanity as virtue. When we recognize the human species as an intrinsic and responsible part of nature, and we feel human identity as part of the wider Earth community, the virtue of humanity naturally extends to other life and to the Earth itself. David Hume, one of the pioneering eighteenth-century philosophers of humanity, saw the "sentiment of humanity" naturally reaching out to other forms of life. He put it very simply about animals: "we are bound by the laws of humanity to give gentle usage to these creatures."[17] We should never be cruel or exploitative in our relations with other life. It was also clear to Hume that there is "a natural progress of human sentiments [and that] the boundaries of justice still grow larger, in proportion to the largeness of men's [sic] views, and the force of their mutual connexions."[18] As humanity connects more intensely with nature and the Earth, our humanity expands and we feel bound to be humane towards the natural environment around us.

Being humanitarian is a moral sentiment that can and must extend beyond the human community. The human subjectivity implicit in the word *humanitarian* does not require a human object. As humans, we can feel beyond ourselves into other things and species. The famous Good Samaritan of Christianity in Luke's Gospel could just as easily have stopped to feel humanity for a wounded animal, a terribly polluted river, a dried-up well or a melting iceberg. In today's Earth emergency, the lawyer's famous question to Jesus in this story—"Who is my neighbour?"—can be answered environmentally as well as humanly. Pope Francis makes this clear: "we can feel the desertification of the soil as a physical ailment and the extinction of a species as a painful disfigurement."[19] Humanitarianism is not only for humans.

## Rewriting the principle of humanity

So, how shall we revise and rewrite the anthropology and intention of humanitarianism's first principle?

I suggest three formulas that offer different options for humanitarians to update their fundamental commitment in the Earth emergency. Each rewrite emphasizes humanitarian action as still grounded primarily in an emergency ethics, and recognizes the mutualism of this emergency for nature and humans. Each one of the three options calibrates this mutualism slightly differently to give varying levels of significance to nature and other life in the practical mission of humanitarian action.

### Revision one

The first revision reads as follows:

> *To alleviate suffering in the Earth community wherever it is found and to protect all life and health, while ensuring the respect and dignity of the human being.*

This revision prioritizes *all life* in humanitarian action: human life, other life and nature, but with the caveat that it does so by respecting human beings, and so could not disrespect and degrade humans in order to prioritize nature. It also adds the notion of human dignity, which is not explicit in the original principle but has since become essential in understandings of human rights and humanitarian aid.

Institutionally and operationally, such a comprehensive expansion of mission and role will be challenging to humanitarian agencies as they are today. It will involve the integration of new ecological and environmental expertise and skills, or close collaboration and mergers with environmental institutions. For example, it would see UN OCHA, UNEP and IUCN working as one in many emergencies to monitor human and non-human need and suffering, and responding to both. In doing so, they would be working jointly with human rights and the rights of nature.

Operationally, an *all-life* purpose will be challenging too. Real life situations are bound to pose questions of operational priority that turn on the doctrine of human uniqueness and different levels of mutualism between humans and nature. Such operational problems arising from an all-life mission can be illustrated in a couple of thought experiments. For example, if a medical team is driving to attend to people suffering from heat-related conditions at a rural public health clinic and they pass a herd of cows which is also evidently suffering from heat stroke and dehydration, should they stop and prioritize the cows? Or, if water trucks are carrying water for people in informal urban settlements and pass a wildfire destroying a maize crop, should they use the water to put out the fire and protect the crop?

Revision two

To reduce such moral clashes, the second revision offers a more human-centred mission, but still values nature and humanity's integral relationship within it:

> *To alleviate suffering wherever it may be found in the Earth community by protecting human life, dignity and health, while ensuring respect for the natural environment and supporting the vital mutualism between humanity and nature.*

This revision seeks to find a better balance between caring for humans and nature, and explicitly emphasizes the need to respect the necessary relationship between them. The word vital is used deliberately, meaning essential to life, as derived from the Latin word for life.

Institutionally, this formula still demands a step change in environmental knowledge and skills, but it is a more explicitly balanced approach which leans more towards human life and thus might not throw up such extreme choices between human life and other life.

## Revision three

The third revision is more clearly and traditionally focused on human life as follows:

> *To alleviate human suffering wherever it may be found in the Earth emergency by protecting and adapting human life and dignity in harmony with nature.*

This minimalist formula stays largely human-centric. Its main change is to expand the humanitarian mission beyond lifesaving to nature-based adaptation. Of the three revisions, this one has the strongest sense of human uniqueness and places less explicit emphasis on the value of other life. Institutionally, it will require better environmental expertise and additional expertise, and strategic collaboration in adaptation.

My preference is for revision two, or something close to it, because it is the most complete and well-balanced formula. It explicitly recognizes the Earth's importance to humanity—an Earthly humanity—and requires a humanitarian response that fosters respectful and harmonious mutualism between humanity and nature.

The last two chapters have made the case for a revision in the fundamental idea and ethical principle of humanity by deepening it to take more account of the embedded relationship that exists between humans and nature. The next chapter explores how impartiality, humanitarianism's second principle, must also change when we find moral and emotional value in nature and see humanitarian fairness in the context of a long emergency that extends significantly into the future.

# 7

## Extending Impartiality

Impartiality is the second foundational value of humanitarianism today. It too has stayed the same since it was conceived and agreed as a guiding principle in 1965. Being impartial, a humanitarian organization:

*Makes no discrimination as to nationality, race, religious beliefs, class or political opinion. It endeavours to relieve the suffering of individuals, being guided solely by their needs, and to give priority to the most urgent cases of distress.*

This gives impartiality four important dimensions: non-discrimination between people; needs-based objectivity in the allocation of aid; a limited ambition of relief; and priority setting on the basis of urgency. Impartiality is essentially the humanitarian's guide to the moral principle of fairness. It sets out a reasonable and fair way to distribute resources in emergencies when they are scarce.[1]

Updating impartiality for the Earth emergency involves significant adjustments around the *scale* and *timeframe* of need, and the legitimate *subjects* of need. The unprecedented and universal impact of climate change and the Earth emergency means we can expect a new scale of human need around the world. The fact that the climate emergency is a long emergency also means we are required to think ahead about the continuation of urgent climate-related need well into the future. Our deepened sense of humanity with nature, and the urgency of saving our natural environment, inevitably broadens our sense of need to nature and other species as rightful subjects of need and recipients of relief. This poses important questions about non-discrimination.

These new ethical demands mean impartiality must be extended sideways to meet bigger human need and to include nature's needs, and extended forwards into the future to take account of the needs of future generations. If Earth system tipping points are triggered into unprecedented, irreversible and cascading environmental changes, the principle of impartiality is likely to face increasing human and nature needs in both directions.

This chapter examines these changes in the scale, subjects and timeframe of need, and also suggests revisions to the principle of impartiality. The first challenge is an extension of human need.

### Massive need

In 2017, the American science journalist, David Wallace-Wells, wrote a powerful essay in *New York Magazine* about the likely impact of climate change this century.[2] Famously, it started: "It is, I promise, worse than you think." Wallace-Wells then described a potentially uninhabitable world—even before tipping points were tipped—with disastrous heat, flooding, food shortage, poisonous oceans and climate-related diseases. Once tipped, he envisaged coastal cities disappearing and 1.8 billion tonnes of $CO_2$ rising into the atmosphere when the world's permafrost finally melts, more than double the $CO_2$ already in the air.

Wallace-Wells felt strongly that "scientific reticence" was presenting overly cautious predictions. He wanted to jolt New Yorkers into action with his finely crafted prose description of what the IPCC's rather numbing technical jargon would actually look like on Earth. His essay is a vivid, well-informed and scary interpretation of the scientific facts and a very important worst-case scenario for humanitarian horizon scanning.

Flying to Jordan in 2023, I read a copy of the Regional Climate Factsheet for the Middle East produced by the Red Cross Red Crescent Climate Centre.[3] Like many international humanitarians, I sat rather hypocritically in a plane that was belting out carbon emissions all around me while I read what these emissions are likely to do to humans and nature thousands of metres below. I expected less brilliant prose and less alarmism from the report of the Red Cross and Red Crescent scientific team. While the prose was certainly more technical, the analysis and prediction were just as worrying. Wallace-Wells was not exaggerating about what might happen.

Average temperatures are consistently going up in the Middle East. Droughts are becoming more common and dust storms will increase. Rainfall is falling across the region. Already, water availability has dropped by seventy-five per cent in the last seventy years. At time of writing, twelve countries in the region are experiencing absolute water scarcity—when water supply drops to 500 cubic metres per person per year. When it does rain, flash floods are worse because so many fast-growing cities have large unplanned areas with exposed and vulnerable informal settlements. Yet, "the current landscape of climate change barely scratches the surface of what is to come" in the region of the world "which will be most affected by climate change and global warming." The Middle East is expected to be the first region to run out of water. As things stand, temperatures are expected to reach 56 Celsius and render the region uninhabitable by 2100.

These climate-related trends are already putting intense pressure on agriculture and livestock production. Rainy seasons, irrigation systems and viable grazing areas are already diminishing. Such livelihood threats will require serious investment in successful economic diversification, water and food security, effective social protection schemes or orderly migration. Even in the medium term, there will be a significant increase in the number of "exceptionally hot days" in the Middle East, which will make life hard and unhealthy in the region's fast-growing cities. These climate risks demand exceptional adaptation in the economy, life-supporting infrastructure and urban planning across the region.

This kind of near-term and medium-term forecasting suggests that during the race between climate change and adaptation, humanitarians should expect to see what we might call *massive needs*. They will be massive because they will occur simultaneously across large areas of the world, not just sporadically in geographic pockets, and they will affect all life.[4] Gradual degradation in the conditions of life and the habitability of many areas will be compounded by recurrent climate-related shocks, like heatwaves, floods and storms. Unless there is a transformative step change in the pace of adaptation and sustainable economic growth, a significant proportion of the world's at-risk population will become poorer and likely to be rendered destitute.

The scale of the climate emergency in the late 2020s and 2030s may lead to a constant cascade of disastrous events that overwhelms the human capacity to cope in many places simultaneously. Many governments may fail to provide a bare minimum of aid for the billions of people in need. The climate emergency could become a continuous global catastrophe rather than a slow sequence of occasional and distinct disasters.

If we take the World Bank's initial estimates of global vulnerability to climate shock that we saw in chapter 2, we can make a guesstimate of the scale of urgent need in the 2020s.[5] Of the 4.5 billion people already exposed to extreme weather, some 2.7 billion of them are low income or very poor people. Even if only half of these people—1.35 billion—get dramatically poorer and destitute because of climate-shocks, a very large number of people will be in need. If most of them slip into serious climate-related poverty and suffering, the challenge will be huge. Either scenario would push emergency demand well beyond the reach of the current Western funded system of humanitarian aid coordinated around the IASC, which is already failing to meet the individual needs of 315 million people in 2024.

Such a constant universal emergency—or total disaster—would present a huge moral shock to humanitarian institutions. It could generate a feeling of helplessness in humanitarians and force them into the tragic realm of triage, in which a simpler form of catastrophe ethics centred on human survival and strict prioritization is demanded of them. This would be in stark contrast to the more utopian ambitions that many humanitarians have today, which seek to improve all areas of human life and human rights to food, health, education, employment, economic security, shelter, gender equality, representation and wellbeing. Such ambition might not be practical at all when faced with massive needs.

Instead, humanitarians may be forced to retreat to a much tighter definition of *lifesaving needs* as suggested in impartiality's limited goal of the *relief* of suffering. Relief would focus on immediate human bodily survival in extremis, and a minimal package of *life-keeping* aid which enables to people to keep going and develop survival capabilities of their own in the medium term. Wider ambitions of helping to provide more holistic and prosperous *life-making* may be hard to justify and achieve. Emergency humanitarian aid in the worst case of climate emergency might be confined to securing people's *vital interests*, rather than their longer-term wants and desires for a comprehensively good life.[6] These vital interests are essential emergency needs which enable us simply to live. In this scenario, humanitarian relief would come closer to the seven corporeal works of mercy in the Catholic tradition: feeding the hungry; sheltering the homeless; clothing the naked; helping the sick; visiting the imprisoned; and burying the dead. Spiritual (or emotional) works of mercy like consoling, comforting and advising would also be important, especially to foster dignity, courage and hope.[7]

With massive needs and huge numbers of people suffering, humanitarianism may need to change its *unit of analysis*. Western humanitarian aid is essentially individualistic. What Joel Glasman describes as "humanitarian needology"[8] counts individuals, one by one, as people in need (PIN) and often delivers to them in lists of individual caseloads of some kind. But helping huge numbers of people in a continuous universal emergency may require a change in the unit of analysis and operational focus from the individual to geographic areas, the functioning of infrastructure and services, or large group allocations of various kinds.

By necessity, aid may have to become less personal and individualistic and more structural and environmental as a landscape approach. Humanitarian allocation would focus on the general provision of water services, food systems, cooling centres, evacuation systems and health systems, instead of individually linked packages of aid. Massive needs would see humanitarians moving away from a retail model to a wholesale model, so streamlining their costs and thinking about large group impact or area effect more than individual input and outcome. AI may help exponentially to spread social protection payments and disaster-related information digitally, able to simultaneously reach millions and billions of people without them meeting a humanitarian worker.

This would reverse the current logic of fine-grained needs assessment into a simpler one of rationing and resource optimization.[9] A traditional humanitarian approach asks: "what do individual people need?" and then tries to fundraise for them as a variety of different caseloads of individuals. A massive needs approach would ask "what budget do we have?" and then try to allocate it for maximum impact. Instead of counting who needs what in long lists of individuals, humanitarians would start with what resources they actually have and invest it in a way that delivers maximum possible return (in lives saved or vulnerability reduced) across a particular landscape.

This rationing approach can still apply the principle of impartiality by focusing on relief, non-discrimination, needs objectivity and urgency, but it will never be as fine-grained as working on individual lives and caseloads, even if it is still informed by qualitative and quantitative data insights into lived experience. And it will not save everyone, but then nor does the traditional approach, which seldom raises what it needs and works with inevitable blind spots.

## Nature's need

The significance of other members of the Earth community is the second sideways extension of scale and need that challenges humanitarian impartiality as it stands today. Nature too may have massive needs. If we accept that we are an integral part of a wider Earth community and that we depend on these other forms of being to survive and live, if we feel that they are like us in this emergency and we feel humanity towards them, then we must include them in humanitarian action. I have done so in my revisions to the principle of humanity and will now argue for a way to do so with impartiality too. Not to include nature in humanitarian action would be an unethical and self-defeating form of discrimination. Unethical because the virtue of humanity tells us that nature and its different forms of life are precious. Self-defeating because we depend on wider life to survive ourselves. The Earth must continue as it is if humans are to continue as they are.

Here, we must be realistic and shape a form of humanitarian action that responds to nature's need and human need together. Humanitarians cannot rush around trying to save every individual animal, reptile, fish and plant at the same time as they are rushing round trying to save every human life. Or can they? Obviously, they could try, but then not much of anything would get done because other members of the Earth community greatly outnumber humans. For example, there are probably 3 trillion trees and 3.5 trillion fish in the world, and vast areas of rice, maize and wheat.

To be practical and realistic, which is important in ethics, impartiality must find a way to meet human and nature's needs together. If humanity is embedded in nature, then nature should be embedded in humanitarian action. Combined programming of some kind, which rightly takes a landscape approach, is the way to find the right balance. Pragmatism will be best achieved by designing humanitarian action that specifically saves and supports the vital mutualism between humanity and nature.

However, the relative mix of a combined approach is bound to change according to circumstance and situation.

In some instances of the Earth emergency, the immediate needs of nature may become much greater than the needs of humans, and so demand that humanitarians focus primarily on saving the lives of other members of the Earth community. This might be especially true in wildfires, or wetland and coastal preservation, or in urgent programming to prevent desertification like Africa's Great Green Wall, in which eleven countries are working to reinforce and expand trees, vegetation and crops before so many species are killed or driven out of the Sahel. At other times, it will be immediate human needs that take precedence. In most humanitarian action, however, I expect that a combined approach—one that meets the needs of many groups within the Earth community—will be both necessary and feasible.

Despite their frustrating annual funding and budgeting cycle, most humanitarian agencies operate with a three to five-year view of climate-related hazards and disasters. Most DRR, CCA and humanitarian emergency responses centre nature-based solutions (NbS) that work this medium-term view to save the life of a variety of species in the Earth community as well as humans. Mitigation efforts of humanitarian agencies to end the use of fossil fuels, like those transitioning to renewable energy, similarly share an all-life goal. These initiatives are saving, seeding, cultivating and protecting a variety of life in their effort to build resilient spaces in which the Earth community can survive and thrive.

A core goal of NbS programming is rightly a net gain for biodiversity, ecosystems and human life.[10] For example, the Honduran Red Cross has reduced the risk of landslides in the pedosphere by increasing slope stabilization with new vegetative planting and organic fascine drainage across degraded slopes. The Vietnamese Red Cross has spent the last thirty years planting and restoring 9000 hectares of mangrove forest in coastal areas. This has restored biodiversity and is also protecting 350,000 people in 100 coastal communities from storms and floods.[11]

This type of combined, or integrated, programming recognizes humanity's relational ontology with nature, and involves humanitarians living out the concept of mutualist humanity and all-life ethics. In Pope Francis' language, this is ecologically integrated humanitarian action and does not discriminate against other members of the Earth community. More poetically, perhaps, it is programming that achieves humanitarian harmony between human needs and nature needs.

## Future need

The second aspect of the Earth emergency to challenge impartiality is future need. Blaise Pascal, the French philosopher famous for his *Pensées*, found humans deeply frustrating because of their constant thinking ahead and their nostalgic remembering of the past.

> We never keep to the present… We are so unwise as to
> wander around in times that do not belong to us, and do
> not think of the only one that does… The past and present
> are our means, and the future alone is our end. Thus we
> never actually live.[12]

Pascal makes an important point about the richness of living in the present, and the dangers of being too dispersed across time. It is a point with which many Buddhists would certainly agree. When dealing with what Thomas Hale calls "long problems", however, we have no choice but to look ahead as carefully as we can. Hale rightly points out that thinking about future generations is "a basic moral intuition" that is found universally in all human cultures.[13] It is right to try to understand, as precisely as possible, how what we know and do in the present affects life and needs in the future. Hale's excellent book, *Long Problems*, argues that, despite the endless distractions of short-termism today, "the objects of governance are increasingly in the future" and "governing across time" is becoming an essential skill for humanity as a whole.[14] This includes humanitarians.

Climate change is a long emergency. We know that it is likely to continue as a life-and-death challenge to humanity and nature for decades. If mitigation and adaptation succeed soon, we may only have to live with what scientists call a "residue" of higher temperatures, changed habitats and biodiversity, which have already happened and are irreversible. The worst tipping points will be avoided. If Earth systems tip, however, then it will be an extreme and very long emergency. We are bound to think about the future in either case, just as we do in any area of life and ethics. We save money for the future. We educate our children so that they can eventually secure good jobs. We plant trees and flowers to have a more beautiful garden in the future.

In the Earth emergency, the future is a legitimate part of the emergency of the present. Pope Francis reminds us that "the notion of the common good also extends to future generations" so that justice itself demands "intergenerational solidarity."[15] Knowing that human distress and nature's suffering is likely to be most acute in ten to twenty years brings urgency about the future to our actions in the present, and must influence the way we spend humanitarian money and time. We feel for the future, and our humanity reaches beyond the present to care about all kinds of future life. This temporal extension of humanity tells us clearly that the future matters ethically to us. In so doing, we recognize the possibility and morality of *intergenerational aid*—doing things now to help life in the future. Some of this life already exists, like today's children, trees and wetlands, which will simply be older in the future, but other future life that matters is still unborn and does not exist. We are duty-bound to take them into account.

Simon Caney, a pioneering philosopher of climate justice, spells out the general duty of intergenerational justice: "the members of one generation should act in such a way that they leave future people with a standard of living that is at least equal to their own."[16] This humanitarian duty becomes especially binding under two conditions, both of which apply in the Earth emergency. Firstly, if the deterioration in conditions looks likely to be catastrophic for the next generation, which it does with climate change. Secondly, if our "temporal position" in this long emergency gives us particular foresight advantage and practical capacity to act successfully in the interests of future generations.[17] In the Earth emergency, this additional foresight obligation to work for the future also applies.

So, despite Pascal's protestations, the present is not everything that matters and the future is an important moral place. As a result, impartiality in a long emergency must think fairly about future needs as well as present needs, which are bound to compete sometimes. This can be especially challenging because thinking into the long future is inevitably imperfect: uncertainty about the future means that we must often act before we know for sure how best to do so. With this in mind, updating impartiality must still involve finding a way to value present and future needs at the same time and with the same budgets, and allocate resources fairly for both.

This temporal competition is always difficult in public policy even when there is no long emergency. For example, deciding to invest significantly in a new hospital which may take five years to build is about improving healthcare for the next thirty years and takes into account many thousands of people who are not yet ill and not yet born. At the same time, however, it ties up money and energy in a forward-looking project that could be spent immediately on increased public health projects for people who are unhealthy now. In an ideal world, or perhaps in Norway and Qatar, we would have money enough for both. But in the world as it is, we do not. Humanitarians in the Earth emergency will, inevitably, have to choose how they balance resources between present and future need: how much they spend to help people in need today, and how much they spend on early warning, DRR and adaptation to reduce future suffering. A revision of the principle of impartiality must recognize this balance somehow.

In thinking forwards, it is important to consider four particular risks we encounter if we lean into the future too much. The most obvious risk of weighting humanitarian investments heavily towards the future is that the present generation may incur an "excessive sacrifice" because their current needs are severely discounted in preference for future needs.[18] This would be unjust in allocational terms. If all spending was invested forwards, it would leave out many who are suffering today.

Second is an epistemic challenge. Investing in the future is epistemically risky because we cannot know for sure what the future looks like, or if future changes and discoveries may make things better regardless of our intergenerational aid. In other words, we face what Hale describes as the "early action paradox": in responding to future crises we are required to act before we know exactly how best to act.[19] Epistemically, we also face what Hale calls the "shadow interests" of future generations.[20] Because the people in future generations do not yet exist, they have no agency and cannot present their interests to us directly in the present. They are shadows on the emergency horizon which we must gauge and interpret as best we can from the present. Epistemic risk also rises over the long term. It is relatively easy for humanitarians to look 3–5 years ahead and see how much worse climate-related living conditions will be then. The communities they work with will also have clear views on the right balance between making a relative sacrifice today to ensure their children's greater resilience tomorrow. Beyond five years, we are probably horizon-scanning, intensifying the early action paradox.

Thirdly, our current institutions are not designed for future challenges. This is Hale's "institutional lag" which slows us down and risks us going off course from the start.[21] Institutions which were well designed as innovations in the past have now become "sticky" and resistant to adaptation towards future problems that need new action now. Their mandates, values, skills and culture do not match the problem coming at us and they are very hard to change. Today, we need agencies that have a combined human and ecological purpose, and well-integrated social and environmental skills. Instead, we mostly have human-centred agencies and environmental agencies working in parallel. And we have a humanitarian bias to war emergency. The war bias means that the whole sector swerves instinctively towards every new war. It might be wiser to let a couple of specialist agencies lead on war while the majority of the sector refocuses on new mandates, principles and practices that match the climate emergency. But old habits and new money are hard to resist, and so there is an institutional lag in climate-related humanitarian change where there should be acceleration.

Some philosophers emphasize a fourth risk too. This one makes no sense to me, but I share it anyway because your mind may well be more subtle than mine and see things that I miss. It is the so-called *non-identity problem* emphasized by the British philosopher Derek Parfit, who was a well-known and very distinctive figure in the streets of Oxford for many years. Parfit's theory undercuts the continuity between present and future generations, and so perhaps our obligations to the future, by saying that the future people (or animals and plants) who we have in mind now as we implement climate action will never actually exist. This is because the actions we take will trigger a chain of reactions, ensuring a very different future population to the one that would have emerged had we done nothing to reduce our emissions. In other words, we will be improving the future for a completely different demographic than the one we care about today.

For Parfit, this means that we cannot truthfully claim to be saving today's future generations at all because that particular generation will never exist now that we have taken them into account. By saving them from terrible lives, those lives will never exist. This deft demographic point seems morally irrelevant to me. As a humanitarian, I am not trying to save particular people or plants or animals in the future, but people and plants and animals *per se*. I do not care *which* people will be alive in the future, but simply *that* they are people, and that they are living better than would have been possible without my intergenerational aid. Some Oxford philosophy can be a little too subtle for a humanitarian like me, but you may get the profundity of Parfit's point and it may stop you in your futuristic tracks.

The moral imperative to invest time and money simultaneously in present and future need requires a temporal revision to impartiality, one that must somehow balance both. Before revising this principle, we also need to explore more precisely the depth of need the humanitarian project intends to address. This hinges on our understanding of another word in the current principle—the idea of *relief*.

## What is relief?

In philosophy, productive discussion and new progress tends to start only when we can agree on the meaning of key words. In humanitarian ethics, especially in discussions about the principle of impartiality, the little word *relief* is one such case. Until we agree what we mean by relief, we cannot set out the extent of our obligations, intentions, and actions as humanitarians in the Earth emergency—either to humans, nature or future generations.

Like many English words, relief actually derives from Latin and French. It is part of Britain's linguistic inheritance from hundreds of years of Roman and Norman colonization. Relief is from a Latin word meaning to "lift up" after someone or something has fallen (*relevare*), or to relieve someone by "lifting off" a burden. Sieges are relieved, just as our financial pressures are eased by tax relief. Using polite archaic English, I might offer to relieve someone of their heavy suitcase by carrying it for them. In this sense, relief is the same as *alleviating*, which defines humanitarian action in the principle of humanity. More specifically, in Norman England the word relief also came to mean assistance. The practice of "poor relief" denotes material assistance given to the poor. These meanings give relief a threefold sense of being lifted up and helped to stand more easily on one's feet again; being freed from pressures and burdens of various kinds; and being given material help. Finally, a physical and emotional sense of relief—in the relieved person—is something we all feel when life is easier again.

When I started working in international emergencies in 1985, I was a relief worker rather than a humanitarian worker. Only in the late 1990s did we Anglo-Saxon relief workers begin to identify as humanitarian workers like our Swiss and French colleagues. Although we all call ourselves humanitarians today, the activities we carry out are still defined quite strictly in humanitarian law and principles as *alleviating*, *protecting*, *ensuring respect* and *relieving*, nothing more. This precise and limited mandate, as it stands today in humanitarian ethics, raises important questions about the range of humanitarian engagement in human life, human society, and the wider Earth community and its many ecosystems. What does relief mean in practice?

Surely, I am doing relief if I help people up from the under the rubble of a storm, or give hungry people food, or heal sick people with medicine and good care. Surely, I also give a wheat field relief if I water it in a drought, or relieve cattle and goats from dehydration by buying them in a destocking programme which relocates them away from a drought-stricken area to a region with plentiful pasture and rain. This kind of relief of immediate needs makes obvious sense. Generally, the words *need* and *relief* make a good pairing for all members of the Earth community.

But what about vulnerability and resilience? Is relief something that addresses entrenched vulnerability of various kinds? Or does reducing vulnerabilities operate in a different realm of structural change, one which involves the elaborate dismantling and transformation of social, economic and political systems that are well beyond relief? Can resilience be increased by relief? Or is resilience-building also a more structural and interventionist process? In the dualism of today, do the reduction of vulnerability and strengthening of resilience fall under development or relief? Are they humanitarian?

How we answer this question determines whether DRR and nature-based solutions are part of humanitarianism or not. If not, then we must say goodbye to a large part of the work of the Red Cross and Red Crescent Movement, which professes to be purely humanitarian, and describe it more honestly as something else beyond the scope of current humanitarian principles.

In fact, relief is clearly a good word which suits many practices in the wide range of activities currently claimed as humanitarian. We can indeed operate generally as "relief workers" because of the flexibility of its different meanings. Relief can be immediate material assistance like food, water and cash, which gets people and nature back on its feet, helping them to stand again. Relief also aptly describes the lifting-off and alleviation of pressures which weigh down on people to make them vulnerable, like the ones in the pressure and release model of DRR we saw in chapter 4, which we could legitimately call the pressure and relief model too. Pulling rubble off people trapped after a storm in a rescue operation relieves the pressure pinning them down, as does redesigning the landscape around them to relieve them from the hazard risks that threaten them, enabling them to stand more firmly to resist a flood, a storm or a heatwave. And the timeframe of such relief does not change its moral purpose or its character as humanitarian. The fact that the Vietnamese Red Cross is taking thirty years to relieve coastal communities of some of their vulnerability to storm and flood does not make it any less of a relief effort, or any less humanitarian just because it is long.

Relief is a wider term than we might initially think because it can involve alleviating all kinds of pressures that put humans and nature at direct risk. It enables people to protect themselves and can also involve giving them things to relieve their immediate suffering. The various meanings of relief also allow us to understand how our inter-generational aid can relieve the pressure on future generations of nature and humans. Relief is a keeper in our revision of impartiality.

## *Revising impartiality*

The discussion above concludes that any new principle of impartiality should be clear about five things: it should lean into potentially massive needs; it should not unfairly discriminate between different human and non-human members of the Earth community; it should recognize the mutualism of humanity and nature; it should take account of life in the future; and it should focus on relief. Two optional revisions may fit this bill:

### Revision one

This first revision offers a general principle that operates with very little detail. It insists that a humanitarian commitment should:

> *Respond fairly to the urgent needs, vulnerabilities and risks of humans beings and other members of the Earth community by relieving their suffering and distress, and protecting their shared environments from current and future damage.*

This option clearly introduces the whole Earth community and recognizes the mutualism of shared environments. It no longer talks of individuals, and is less specific about potential types of discrimination within the human population—nationality, race, religion, class and politics. This revision also keeps the word "distress" because it may not make sense to attribute suffering to all forms of nature.

### Revision two

The second revision is more human-centric and specific. A humanitarian commitment should:

> *Respond fairly to the urgent needs and risks of all human beings and Earth's ecosystems in which humanity shares vital*

*mutual interests, in order to relieve the suffering and distress*
*of vulnerable mixed communities of humanity and diverse*
*life, and protect them from immediate and future damage*
*and harm.*

This revision does not use the idea of the Earth community, but talks instead of Earth's ecosystems, mixed communities and diverse life. It also adds the idea of harm as well as damage.

Both these revisions may work well as first principles from which one can infer more specific secondary principles and duties. But it is also possible to be even more detailed and address distinctions in both humanity and nature that have important implications for social and environmental fairness.

## Human differences and natural enemies

We know humanity itself comes in many forms, and that human experience can be positively and negatively affected by social norms and politics. For example, my experience of fairness in life and climate-related disasters may be different if I am a brown person, a black person or a white person; if I am a Christian person in a majority Muslim country or a Muslim person in a majority Christian country; if I am a poor person or a rich person. It is these aspects of contingent intra-human distinctions—nationality, race, religion, class and political opinion—that the original principle of impartiality seeks to capture and discount in its rule of fair treatment.

Since the principle was agreed in 1965, people have gone further than this list of social differences to explicitly identify gender, age and ability as salient human identities that regularly meet with discrimination and unfair treatment. Thus, it may make a big difference to my suffering and life chances if I am a woman or a man in a climate-related disaster, old or young, or a person living with disability.

We could add these three differences of gender, age and disability explicitly into any revision of impartiality. Each one of them has rightly been given considerable significance in many secondary humanitarian duties inferred from the primary principle since it was agreed in 1965. There are now major humanitarian policies addressing gender, age and ability, which bind humanitarians to firm duties of impartiality, and usually prioritization, around each of these human differences. In the last fifty years, there have also been major new human rights treaties which have outlawed discrimination on the basis of gender, age and ability, and given legal guarantees for the equal rights of women, children and people with disabilities. Only older people are still waiting for a treaty of their own. In a few years we might have to add and include autonomous robotic life into our ethics of non-discrimination too, depending on how conscious and life-like AI becomes.

There is a growing research and policy literature that establishes how women, girls, boys and older people are differently, and sometimes disproportionately, affected by climate change and disasters. Different racial, ethnic and transgender identities may also contribute to a more severe experience of climate change and climate-related shocks. The same is true of older people and people with disabilities. A revised principle of impartiality could find a way to recognize them all as especially vulnerable and at risk because of discrimination of various kinds. However, it is hard to do so without emerging with a long list as well as a principle. The third revision tries to make the point without the list. In so doing, I leave important work to be done by the clarification of secondary duties inferred from the general principle. Establishing new duties and rules from the evolving interpretation of general principles is normal and creative, and usually done in specific policy or standards.

There are also important differences within nature too, which it might be wise to reference in the revision of impartiality. In short, we have enemies as well as allies in nature and the ecosystems in which we live. Should we extend our humanity to them and be careful not to discriminate against them as we save all life? Or can we say explicitly that they do not count and should exist beyond our care as a danger to humanity and one of the risks we want to reduce? I am, of course, thinking largely of our current predators. These are the many microbes who can kill us as bacteria and viruses, as well as the vectors that carry them around, including mosquitoes, rats and plants. It is more truthful to say that we do not value all life and would gladly see these killed or dramatically reduced. We certainly invest a lot of time and money trying to kill them already. I think most of us would agree that we should make exceptions to non-discrimination here. So, humanitarians can actively be involved in killing and reducing certain species to relieve suffering. I think many other species would agree with this because microbes hurt and kill them too.

However, there is still the question of killing and eating. Many species kill other species, usually for food. Our human role as predators needs to be considered in the ethics of impartiality, especially the fact that we specifically rear animals for killing and eat animals of many kinds. Shall we make an exception of our killing and eating in humanitarian ethics and claim this is not unfair discrimination? And shall we ignore the killing-to-eat of other animals and insects too, even when they kill or eat us?

I'm afraid I am going to disappoint vegans and vegetarians by making this exception. I still eat meat on special occasions, and do not yet think I am wrong to do so, or that I am being atrociously discriminating and unfair. One argument I can make for this exception is that of necessity during humanity's transition to a more renewable global economy with less intensive animal farming. The world human population must still eat animals to secure the protein and energy we need to live well until a full system of protein replacements can be universally developed with renewable energy.[22] So, reluctantly perhaps, I decide that killing to eat is not always unfair discrimination—although cruelly keeping and killing animals is inhumane and unjust discrimination while they live.

A second argument is more to do with deep ecology and a sense of reality about life on Earth. There is something ecologically inevitable and justifiable about killing and eating other life across the Earth community as it has emerged and evolved. I cannot require eagles and tuna to become vegetarian. I could domesticate them in certain ways so that they have no choice but only to eat lentils, or I could render both species extinct because I deem them wrong for eating others. However, as part of the Earth community, I find I can still respect and thank the animals I eat, while recognizing that various life forms will probably kill me one day too, and eat me if I am buried. The Earth community interacts in killing and death as well as in cultivation and life. Our relational ontologies involve ending life as well as cultivating it.

### Revision three

A third revision, which takes more account of differences in humanity and nature, goes like this. A humanitarian agency should:

*Respond fairly to urgent present and future needs of humans and nature, without social or political discrimination between people. It will endeavour to relieve the suffering, distress and risk of human communities and the ecosystems on which humanity depends, prioritizing on the basis of differences in human vulnerability and capacity, and the positive contributions of diverse life.*

This formula is long, but captures differences in human vulnerability and nature's contribution, and prioritizes accordingly. It does not use such personal language about the Earth community or make reference to individual humans (human beings), but focuses instead on larger human groups (human communities) and ecosystems as its units of analysis.

All three of these revisions to impartiality, and the three earlier revisions to humanity, insist on the inclusion of the wider Earth community in humanitarian action. This means, of course, that humanitarian needs assessment must start analysing distress, vulnerability, capacity and risk in nature, and humanitarian programming must plan, budget, deliver, monitor and report humanitarian impact in the natural environment.

This chapter now completes the possible revisions to humanitarianism's two most significant and fundamental principles—humanity and impartiality. These are the basic values of humanitarianism. As their original drafter, Jean Pictet, famously observed, they are the *substantive principles* which "inspire" all humanitarian acts and so "belong to the domain of objectives".[23] The next two chapters will explore new features of the Earth emergency which merit new recognition as valuable and action-guiding operational principles for humanitarianism in the Earth emergency, starting with anticipation.

# 8

# Embracing Anticipation

Human beings have always tried to anticipate the weather, and tended to do so in two main ways. Firstly, by remembering weather patterns from the past in order to interpret the present. Farmers constantly recall certain characteristics of previous summers, autumns and winters to match them with the present, and so predict the likely temper of the planting and growing seasons immediately ahead. Secondly, humans instinctively observe animal and plant behaviour in the present to seek clues about what is coming. We watch animal activity and plants blossoming to detect when it is about to rain, or to see if spring is arriving early or delayed, and if our corner of the Earth is warming or stubbornly staying cold. This ancient and intuitive pursuit is known scientifically today as phenology. Seeing or hearing bird behaviour is often its easiest indicator.

In England, as our side of the Earth tilts again towards the sun at the end of winter, we hear birds singing different songs, and see them forming pairs and gathering material to make new nests. Migratory birds also arrive again as England starts to warm and the days begin to lengthen. My face lights up when I see swallows and swifts for the first time each year, watching them swirling high in the sky, or swooping in under the roofs of our houses to rebuild their nests from last year. I especially love the fact that they may have been somewhere in Africa over the winter and are bringing a bit of African style back to the dreary old UK.

I am in awe of the fact that swallows have flown thousands of miles to get here, with many of them returning to exactly the same place. My little house usually has two swallows' nests. As they fill up again each year and hatch their young, the house fills with the noise of their bustling, feeding and tweeting, and we live again as a mixed community of humans and swallows under the same roof. Early spring is a time of great anticipation not only because of busier birds. We can also look forward to being warmer and surrounded by much more life again, which is singing, buzzing, budding, blooming and ripening around us.

Living in the early stages of the climate emergency, we must also embrace anticipation as part of the new ethical paradigm for humanitarian action. We know that "caring for ecosystems requires far-sightedness" and that without ecosystems humanity cannot survive or thrive.[1] In a rapidly changing climate, however, only some weather and hazard patterns will be recognizable from the past. Much that happens will be unprecedented and we will be left unguided by retrospective analysis of the past. This new era requires scientific expertise in anticipation and prospective insight as we assess what nature's behaviour in the present might tell us about future weather. Only by genuinely exploring the future, and not simply reading the past in search of a similar future, will humanitarians be able to prepare for and respond to the impact of a dramatically changing climate on humanity and nature.

Thanks to modern science and its improved data collection and analytics, we are able to anticipate natural phenomena and their likely impact more precisely today than ever before in certain areas of Earth systems monitoring. This enables us to anticipate the suffering and damage that is to come. Our revised impartiality principle affirms that we should look ahead to relieve suffering, and to a significant degree we can. Therefore, anticipation becomes an important humanitarian principle in this long Earth emergency.

## Near-term anticipation

The predictions of the IPCC give humanitarians strategic foresight into what we can expect in the next few years and decades. This medium- and long-term view demands a reduction of greenhouse gas emissions and intense efforts in whole society adaptation over many years. This chapter focuses on advances in short-term forecasting and more immediate anticipation, which is different from long-term strategic foresight.

Scientific progress in anticipation places a new moral demand on humanitarians and all human communities to be alert to specific climate-related warnings and agile enough to respond positively to them, in order to prevent loss, damage and harm they know is likely to happen in the next few days, weeks and months. This near-term perspective is intragenerational rather than inter-generational, but it still carries some of the ethical challenges of moral foresight and the inevitable trade-offs between resource allocation for present or future needs.

New anticipatory capacity in reliable near-term forecasting is enabling humanitarians to see ahead in the short term. This near-term insight gives humanitarians new scope for preventive action and imposes an obligation to anticipate the immediate future and act wisely to protect people and nature in advance. Much of this near-term anticipation is effective only in days and weeks. For example, detailed scientific weather *forecasting* is 80% reliable for up to seven days in advance and 50% reliable up to fourteen days. New abilities in *nowcasting* mean that accurate weather forecasting can be made for the next six hours for highly localized areas. This very near-time forecasting window can be vital for specific humanitarian preparations, like evacuations, cooling centres, medical services, food supply, and cash services in advance of storms, floods and heatwaves.

*Seasonal forecasting*, which combines dynamical modelling from current conditions with analogical historical data, is also improving and able to inform agricultural and other planning. These forecasts give probabilistic average estimates for seasonal conditions over increasingly large areas and are framed in language like "there is an increased chance", or seasonal rainfall "is expected to be", and so on.[2] These forecasts are becoming increasingly good at spotting extreme climate patterns likely to cause drought across East Africa, which researchers now feel confident that they can predict eight months in advance.[3] Meteorology is also improving its ability to detect and predict large medium-term weather trends like the El Niño effect, which is having a sustained impact on global weather as I write in early 2024.

This new sophistication in dynamic atmospheric measurement, monitoring and modelling, which is increasingly using AI, is dramatically improving near-term weather prediction and will continue to do so. This gives us new levels of insight and certainty in *hazard forecasting*. Humanitarians are also better placed to anticipate what will happen to people and nature in these foreseeable climate-related shocks because we also have a better grasp of *impact-based forecasting*. By looking back at what happened to people and nature in similar hazard events, and using AI to model this data forwards, humanitarians are able to improve forward-looking estimates of what losses, harm and damage can be expected, and so gauge the particular risks people face. For example, these impact-based forecasts may expect a 25% loss of housing from floods, a 30% loss in crop yields from extreme heat, or a 50% increase in a particular climate-related diseases like malaria, dengue or heatstroke. This kind of prospective insight allows humanitarian to respond to needs in advance.

The difference between hazard and impact forecasting is well described by the UK Met Office and the Red Cross and Red Crescent Climate Centre: hazard forecasting predicts what the weather *will be*, and impact forecasting predicts what the weather *will do*.[4] Both types of forecast are important components in *early warning,* and together they now enable the possibility of *risk-informed early action*. There is evidence that timely, effective early warning and early action can reduce disaster mortality to eight times lower than no early warning, and reduce damage by 30%.[5] This is why three UN agencies and the IFRC have come together in a global initiative for Early Warning for All (EW4A) to ensure global coverage of early warning systems that use forecasting of all kinds.

For an early warning system to be effective, it needs to focus hard on four areas, which are the "four pillars" of EW4A. Firstly, communities and governments need to work out the risks they face from different hazards and get to know them well. Secondly, the hazards must be detected, monitored and forecasted by satellites, computers and scientific analysts. Thirdly, warnings must be communicated in an accessible way to the right people at the right time. Finally, governments, people and humanitarian agencies must have the plans, preparedness and resources in place to respond fast and appropriately to the warnings in order to save life and reduce damage.

This last point means that early warning systems must be linked to quick and *shock-responsive action* to move people, animals and assets to safety, and supply them with relief that is timely and relevant. Innovative forms of *disaster risk financing* are being increasingly used to finance such humanitarian action, aiming to limit climate shocks and hazards that may either be slow onset, like a creeping drought, or sudden onset like a heatwave, flood or storm. Disaster risk financing includes an array of programming like special disaster insurance, rapid cash transfers or shock-related increases in existing social protection schemes. These various anticipatory schemes use *triggers* of various kinds, which are pre-agreed with all parties to enact aid when warnings crossing a certain threshold.

The German Red Cross has been a major pioneer of anticipatory humanitarian action and established the Anticipation Hub in Berlin in 2020. Anticipatory action around the world is still small, with only 70 schemes in 35 countries reaching 7.7 million people in 2022.[6] The UN's

World Food Programme (WFP) is prioritizing anticipatory action around food security and aiming to support 5 million people in such schemes in 40 countries by 2025.[7] Much is being learnt in the early pilots to-date, especially those working in tandem with insurance companies. Building on earlier innovations by Oxfam, WFP is now partnering with governments of high-risk countries to expand national insurance schemes at a macro- and meso-level. These schemes see governments paying regular disaster insurance premiums to guarantee a rapid response from WFP in cash or food relief when warnings are triggered. These interventions are then covered by the insurer.[8] The START Network is specializing hard in locally led planning and response to prove the case for localized anticipatory action.[9] In 2019, a coalition of governments, the IFRC and humanitarian agencies launched the Risk-Informed Early Action Partnership (REAP) which aims to reach 1 billion people with effective early warning and relief response by 2025. The race is on.

Preventive action merits expansion by governments and humanitarian agencies for hazard types and emergencies that are amenable to near-term prediction. Such preventive programming for disasters is analogous to vaccination in public health policy. However, while humanitarian agencies are right to adopt a new paradigm of anticipation in their work and ethics, they should not become completely mesmerized by it as an easy or singular solution. The widespread adoption of anticipation will involve some operational traps that pose particular moral risks. Before scrutinizing this shadow side of anticipation, it is important to argue the bright side first and explain why anticipatory action is both ethically important and a clear moral good.

## Anticipation as moral value

As you might expect, there is a large academic literature on anticipation. Every discipline is studying if and how humans, nature, complex systems and computers can think and see ahead.[10] In an era dominated by an unprecedented number of existential risks, the intellectual hive is reorientating towards the future.

Italian sociologist Roberto Poli points out that the future is increasingly becoming "a core organising principle of the mind" as every discipline and profession today is forced to look forwards more than backwards. This means "turning the humanities and social sciences upside down".[11] Instead of seeing the past as primarily determinative in human affairs and the main cause of how things are today, most professional fields now see historical trends as increasingly irrelevant and recognize causality as better-positioned in the future. In other words, what we do in the present should not be so firmly guided by an appreciation of the past, driven instead by our anticipation of the future before us. In the climate emergency, we urgently need to be changed by the future rather than the past. This temporal reorientation puts anticipation at the centre of human concern.

Anticipation theory was launched out of the natural sciences in 1985 by the American biologist Robert Rosen, who observed "the ubiquity of anticipation" across the natural world in his book, *Anticipatory Systems*, and developed a theory which became the conceptual basis for anticipation and foresight studies.[12] The main difference between a reactive system and an anticipatory system is the key to Rosen's paradigm shift. A reactive system can only react in the present to changes that have already occurred, while an anticipatory system's behaviour in the present uses aspects of the past, present and future interpreted by an internal predictive model of some kind. As a result, anticipatory systems use *feedforward* more than *feedback*. If feedback is a form of correction to the future based on past experience, feedforward uses expectations of the future to change the present.

The distinction between reactive and anticipatory systems turns mainly on a different sense of time in an anticipatory system. Temporality is more open and fluid in an anticipatory mindset. Rosen, and other biologists, have observed all varieties of life behaving with the future in mind somehow, as if their inner model of the future gave them "an internal surrogate of time", which runs faster than real time so that they can change in time for the future. This process of getting ahead of time in order to anticipate it and respond to it in advance is, of course, exactly what humanitarian anticipatory action tries to do with its (inner) models of hazard and impact forecasting.

Like everything in nature, humans are anticipatory too, which takes us back to Pascal's complaint that we never live simply in the present. In our personal lives and in our public policy we constantly build anticipatory systems. We care simultaneously about things past, present and future. This is an ethically good thing even if it is a spiritually distracting thing, and sees you worrying about a work deadline when you are meant to be experiencing the perfect present in a mindfulness session.

Acknowledging anticipation as a part of life is also realistic about how time actually works. Several early twentieth-century philosophers, like Rudolph Husserl and Alfred Whitehead, usefully remind us that the present is actually quite a crowded place, and that *now* is not as simple and singular a moment as it sounds. Husserl held that human consciousness never exists solely in a pure present. In his jargon, there are always three aspects to the present: the retentional, the impressional and the protentional.[13] At any moment we are simultaneously aware of the past, impressed by the present, and expect the future. Such is the threefold dimension of human consciousness. In a similar way, Alfred Whitehead described reality as a process that is always *becoming* more than it is simply *being*. The present is in flux and unfolding. Like William James, Whitehead understood consciousness as a stream, in which any moment is in motion and internally related with other events across time.[14] To use a favourite metaphor of humanitarian bureaucracies, time does not operate in siloes.

This understanding of anticipatory life and fluid time makes it inevitable and important that humans, and humanitarians, are anticipatory. Anticipation makes biological, psychological and moral sense. Biologically, anticipation is a major evolutionary trait that we share with all life and is an essential part of survival. Psychologically, our seamless sense of time explains why, as humans, we feel the future as much as the present and the past, and constantly use the future in our decision-making and our social construction of reality. The year planner on the wall in front of me is an obvious example.

Morally, our anticipatory instincts and our fluid feeling of continuous time explains why we want to deploy the virtue of humanity forwards through time. Our moral sense tells us that the future is already here somehow because the consequences of our actions or inaction now will matter. This sense of consequentialism sits deep within us as moral beings so that what we do in the present is already part of the future, and we cannot ignore the future when deciding what is right and wrong.

This gives significant moral value to anticipatory actions like exploring the future, expecting certain things in the future, and responding to them in advance. Anticipation is good humanitarian ethics even when we do not know exactly what the future will be. The other way we know that anticipating the future must be right is our sense of moral failure when we do not think ahead, and bad things happen which could have been avoided. Then we can be fairly judged as having been thoughtless, uncaring and negligent about what might happen in the future. This means that anticipation should be recognized as an important guiding principle of humanitarian action.

## Moral problems in anticipation

The section above argues that anticipation is a good thing and an inherently human thing that is an essential part of our humanity. However, as with many good things, anticipation is not always morally simple to apply. Like other areas of humanitarian aid, anticipatory action carries ethical risks around collective action, equity and access, competing needs, and the moral hazard of system habits, vested interests and community dependency. Under our revisions to humanity and impartiality, it also runs the risk of human bias. I will take these anticipation problems one by one.

Anticipatory action is socially and organizationally elaborate. Early warning systems and early action relief require intricate and extensive cooperation between financial investors, tech companies, governments, at-risk communities, meteorologists, data analysts and communications teams at a global, regional, national, district and community level, as well as interactions between Earth and outer space, and between humans and AI. Early warning and early response are typically the result of cooperation between dozens of organizations and many thousands of people in one long value chain.

This inter-agency and multi-level mix poses the moral challenge of *collective action* in which every link in the chain is reliant on others for success. It means anticipatory action is a major ethical test of good cooperation. Success turns on the veracity and honesty of data generation—is data well selected, well collected and correct? It also poses questions of fairness in the weighting given to different kinds of data. Is "unscientific" data, based on local knowledge and experience, included and weighed fairly in the system to provide nuance to data gathered thousands of miles away by people who have never visited the country in question, and have probably never been flooded or hungry? Are vested interests held in check so that preferred outcomes of certain actors are not biasing the system? Are free-riders doing and paying nothing and unfairly getting everything they need for free? The other notorious collective action problem is that collaboration between people can only ever go as fast as its slowest member. So, is the anticipatory system being unfairly and dangerously slowed down for no good reason? These system management problems are all ethical problems too, and turn on the ownership, design, organization, tempo and benefits of early warning and anticipatory action. As a collective action challenge, anticipatory action requires regular ethical scrutiny to ensure truthfulness, timeliness and cooperation.

Another important part of the ethical ambition of early warning for all is the "for all" dimension of this goal. Its universal requirement makes it a serious matter of impartiality that raises routine humanitarian questions of *equity and access*. Especially important in the practice of humanitarian warnings is fair communication and linguistic justice. This turns on translation into all necessary languages and dissemination in accessible formats—whether these be word of mouth, radio, television, social media or print. Vital here is that information, warnings and wider climate services come from people, and in formats, that are clear, credible, and trusted by people who need them.

Fair coverage is also essential for a universal system of anticipatory action. This means reaching people with the right form of communication and the right relief response at the right time. This poses a significant last mile challenge. A particularly important ingredient in equity and access is involving people's own *agency* in designing and delivering communication and relief. That warnings and relief are locally led is integral to ethical anticipatory action. When this is not possible, it must at least involve people at risk.

The next ethical risk in anticipatory action is the moral hazard of incentivizing *dependency and non-adaptation*. Creating a slothful and disempowering dependency on aid is an archetypal fear of humanitarians.[15] As a result, a humanitarian commitment "to avoid long-term beneficiary dependence on external aid" and instead to "help to create sustainable lifestyles" sits at the heart of Article 8 of the Code of Conduct for the International Red Cross and Red Crescent Movement and International NGOs in Disaster Relief, and is implicit in the Humanitarian Charter's commitment to enabling a "life with dignity".[16]

A paradoxical problem with predictable humanitarian aid is its predictability. If people and local government come to rely on regular relief whenever certain triggers are crossed in their environment and livelihoods, they may have no incentive to adapt to climate change. The repeated safety net of anticipatory action in the same communities may insinuate a culture of constant coping and never adapting. Such a scenario would mean that anticipatory aid becomes harmful to wider climate action goals and renders communities more vulnerable in the long term if their way of life is ultimately unsustainable and their environment increasingly uninhabitable.

This risk of creating a chronic coping culture, which is non-adaptive, disempowering and ultimately doomed, is clearly already felt by humanitarian agencies and must be taken seriously in anticipatory ethics. This particular moral hazard will always exist unless repeated anticipatory action deliberately cultivates people's agency as a key characteristic of how it works, and firmly links them into wider adaptation programmes.

Humanitarians face another moral hazard in a *system dependency* on anticipatory action that distracts them from other ways of working. If agencies become too invested in this approach to relief in a rapidly changing climate emergency, anticipatory action may pose a *path dependency problem* in which a system is stuck on a path of action that is no longer fit for purpose. Ethically, this means being stuck doing something that may be a good thing but is no longer the best thing. For example, I might be a doctor who is prescribing antibiotics at a time of increased microbial resistance to antibiotics, even when I know that they are less likely to work in each case and that by continuing to prescribe them I am contributing to more widespread resistance.

Anticipatory action works extremely well in particular places and in certain conditions, like imminent floods, heatwaves, storms and droughts. In these situations, it can support an important segment of the world population whose life conditions and way of life are routinely disrupted but not completely devastated by climate-related hazards. This demographic of vulnerable but semi-resilient people may well amount to the 1 billion people whom REAP intend to reach. But the climate emergency will see people's conditions change rapidly and exponentially. In a new situation, the relative stability of anticipatory action's ideal demographic may vanish, and anticipatory action may no longer be feasible or effective at meeting new needs. In short, predictable aid may not work best in unprecedented and unpredictable emergencies.

For example, a coastal community of 500,000 people in Country X may currently fall into the sweet spot of anticipatory action. This population is well organized and mostly living well enough from year to year. With local government and local NGOs, they have good disaster risk finance and other DRR measures in place, and draw down on it for around 50,000 people every year. But two scenarios could very quickly render this kind of risk-informed anticipatory aid redundant, and see its methods fall behind the curve of a wider crisis.

In one scenario, this coastal population's whole area could be hit by two unusually intense cyclones in quick succession, devastating the assets and livelihoods of around 400,000 people in ways that are well beyond the planned corrections of their anticipatory aid. In this scenario, the anticipatory action currently in place is likely to be unfit for purpose, presenting urgent new moral choices for humanitarians about how best to meet their needs.

In a second scenario, life continues well on the coast but drought makes the interior of the country increasingly uninhabitable. Two devastating drought years in succession see a sudden mass movement of 1.4 million people to the coast, where they settle near main roads to be near government services and towns. This new population effectively crowds out the community protected by anticipatory action. Demand for basic goods and services is overwhelming, and causes inflation and shortages of vital commodities. Conditions for these displaced people is close to destitution, and their extra needs are degrading the life and livelihood of the coastal community. In this scenario, while existing anticipatory protocols could be deployed to help the coastal community, the prioritizing of aid to this group may no longer meet an impartiality test if the new arrivals are in much greater need. It might then be unfair to use resources from the pre-arranged anticipatory scheme on the coastal community first.

The main point here is that current models of anticipatory action may have a relatively quick expiry date in a fast-moving climate emergency full of unprecedented events. Finely targeted anticipatory action might not keep pace with changing needs, and might face severe moral competition for the resources tied up in anticipatory systems. If humanitarians are locked into these anticipatory models by system habit or institutional vested interests, then people in greater need may suffer from these agencies' path dependency problem. Past choices would limit present and future actions because they have institutionalized ways of responding that are resistant to change and unfit for purpose in a new situation. In this case, anticipatory action would no longer be truly anticipatory; instead, it would be a historically beneficial practice that is now blocking more appropriate relief.

The twin risks of community dependency and institutional dependency mean that *personal moral judgement* must remain alive and decisive in the management of anticipatory action. Data may only offer a partial view of emerging situations and it is for human judgement, not pre-agreed threshold triggers, to resolve new questions of competing needs unseen by anticipation. If a system is designed to focus on 500,000 vulnerable people and suddenly there are 2 million, there must be humans in the loop to make new ethical judgements on what is right. Similarly, only people will be able to judge if communities are slipping into a culture of coping, not adapting, and being drained of agency by repeated anticipatory aid.

Finally, there will be moral difficulty in finding a balance in relief allocations between human needs and nature's needs, which may tend to manifest in *human bias*. A new moral paradigm, in which impartiality recognizes human and non-human need in a combined approach to saving life, is ethically challenging in action. This is especially true when people can voice their needs but nature cannot speak in ways we easily understand. Sometimes, the cry of the poor may inevitably resonate more emotionally and more politically than the cry of the Earth. At other times, the cry of the poor may be raised in a cry for the Earth as people prioritize nature around them and accept short-term sacrifices as they do so. Coastal communities may prefer humanitarian investments in wetlands rather than clinics; farmers may prefer spending shared between them and their livestock; an urban community may prioritize investment in nature-based cooling rather than immediate air conditioning.

The ideal of an integrated approach to human needs and nature needs, which rightly sees life as a mixed community of mutual needs, may not always hold. Sometimes it should not hold, and we should clearly prioritize saving human life in extremis. Regardless of our love of and need for the Earth, our love for one another must take precedence in certain situations. This is really for three reasons.

Firstly, because our relational bonds within the same species are more intense and morally real. They press upon us in the virtue of humanity and in the genetics of survival, cooperation and communal progress as a group. Because we live as humans, we are closer to humans. Even the most ardent deep ecologists and green colonialists who fight for nature over humans do so in close-knit human groups. Secondly, our sense of a future means that we can accept tragic natural losses with the conviction that, in some way, we can and must restore them and correct them by re-seeding and recreating in a time to come. This commitment to restoration is also important when dealing with the loss of cultural heritage, as we will see in the next chapter. Thirdly, we know that we will suffer more pain from the loss of people close to us if we feel that we have deliberately let them die and could have saved them. This is not to discount the pain of a lost forest or herd, but is simply to recognize that it will be harder to forgive ourselves for abandoning a person rather than a tree. We will be more morally haunted by prioritizing nature than by prioritizing humans. We must anticipate this when we make aid choices in anticipatory action.

This chapter has recognized anticipation as a new ethical principle in a moral paradigm of humanitarian aid which takes future needs seriously and counts them as present priorities. The following chapters look at three other features of the Earth emergency which require a similar elevation of moral status in an updated humanitarian ethics: loss, mobility and adaptation.

# 9

## Recognizing Loss

The climate emergency is demanding big changes from many people around the world and will continue to do so. Heat, flood, wind, rain, fire and smoke will alter our conditions of life, our way of life and the location of our lives. Humanitarians are already meeting millions of people in very difficult moments when their lives and environments are changing dramatically, and sometimes catastrophically, because of the weather around them. This may soon rise to billions of people who are enduring significant loss, deciding to move and struggling to adapt.

Loss, mobility and adaptation have always been integral to human life. They are not new to humanity or to humanitarians. Disaster and war always bring all three of them in their wake. Today, however, they are framed explicitly as key elements of climate justice and global policy. This means that humanitarians must be clear about how we value loss, mobility and adaptation. Should they now, somehow, be action-guiding principles in humanitarian work? This chapter starts with an examination of loss.

## Describing loss

Last week, I listened to a British farmer from Lincolnshire on the radio whose long-held family farm has been under water for the last six months. His family can only reach their front door by boat, and they have lost two seasons' crops. Unless there is massive government investment in new dykes, this family's way of life and place of life is finished. Even if new dykes are built, this stoic farmer recognized that the very nature of climate change makes it unlikely that they will last long because more unprecedented storms and inundation will defy the imagination and budgets of the local planners. It looks as if this family's long tradition of farming in Lincolnshire is over, their assets are worthless and they will have to move and start again somewhere.

Climate justice and climate action focus hard on the "loss and damage" caused by anthropogenic climate change. Losses are hard to predict, but a detailed study of subnational loss and damage by the Potsdam Institute for Climate Impact Research estimates that economic losses from climate change will see the whole world economy experience an income reduction of 19% in the next 26 years regardless of future emission policies.[1] Investing to recover from that level of loss, and securing the necessary green growth to do so, is a huge challenge.

In most places so far, the really dangerous impact of climate change is not instant death from climate-related disasters but extensive and incremental impoverishment because of a loss of assets, livelihood and place. The maverick Danish political scientist, Bjorn Lomborg, routinely celebrates on social media that there is no cause for alarm because so few people are dying in climate-related disasters today, and merrily shows lots of graphs of declining mortality to prove it. His mortality graphs are happily right but, of course, he is missing the point. Yes, thankfully, direct deaths are massively down compared, for example, to the Yangzi floods of 1931 or the Bangladeshi cyclone of 1991. But the economic cost of disasters is huge and rising. Hurricane Katrina in New Orleans saw economic losses of $187.3 billion in 2005; economic losses from increasing storms, floods and fires across the US and many other countries are accumulating exponentially year on year. In its wake, such recurrent economic loss will cause impoverishment, destitution and, ultimately, a significant rise in indirect deaths from climate change.

Climate policy recognizes two main types of loss and damage: tangible and intangible loss, or economic and non-economic loss (as COP documents prefer). Some losses come in a hard economic form of assets, jobs and income like this farmer's land, crops, business and home. Others are more social, cultural, relational and emotional, like the ties he and his family have to their landscape, local community and profession which give them an experience of beauty, belonging, identity, friends and social purpose. And, of course, both intersect. If I lose my home or my work, I am poorer and also feel sadness, pain and isolation. In thinking about loss and damage, therefore, we are simultaneously in the financial economy and the emotional economy. We are also inevitably pondering the moral economy as we weigh what is fair and unfair about such losses, and what possible remedies should be applied.

Jon Barnet and his geographer colleagues have usefully sketched out "a science of loss" which maps out the various things for which people rightly grieve in climate-related loss.[2] They define loss as instances "when people are dispossessed of things they value and for which there are no commensurable substitutes." This makes the important point that true losses can never really be repaired or replaced. There is no substitute for them. A person killed, a place destroyed and a community dispersed are gone. We know this too from our earliest memories, when the toy or pet we loved was lost and the new one was simply not the same. There is no real and precise replacement for a true loss, just new things to incorporate into our life to give it value once again as we try hard to keep going despite our loss. Loss means something is gone forever; but some kinds of damage can be repaired to be as good, or better, than before.

So how shall we understand the value of deep losses arising from climate change? Barnet's team sees value being lost in two main ways: the loss of *primary goods* like safety, health, income, freedom, belonging and esteem; and the loss of *phenomena that constitute meaning*, like landscape, sacred places, particular roles, jobs and occupations, distinct cultures, and a social cohesion that is born of all these things, and which ends when they end.

Heat, pollution and new climate-related diseases can take away a person's health and compromise their way of life and their possibilities of change. Climatic conditions can hem us in and reduce our freedom and damage our health. For example, millions of people face restricted movement because of extreme heat and its accompanying pollution in Delhi, Dhaka, London, Lahore and many other industrial cities that are heating up fast.

Losing a landscape forever to drought, inundation, storm damage or fire ends assets, livelihoods, community and personal identity. It is a loss of goods and meaning. Billions of people have very sophisticated knowledge, skills and expertise which relate to a particular place or occupation. When this place or its professions are gone, like farming in arid lands, fishing over coral reefs or mining coal, people can be left with finely tuned skills that no longer have a practice, like the phantom pains in the bereft nerves of an amputated arm.

## *Living with and beyond loss*

While we can gauge the value of things lost in the climate emergency, we must also admit that loss is not unusual in human life and society. Nobody goes through a life without loss, pain, scars and change. A big part of being human is losing but still living, and being determined to recover a sense of meaning, joy and purpose after loss. Billions of people are doing this every day around the world and always have.

But there is more to loss than the everyday losses we all share. There are also periods of catastrophic collective loss. Climate change is by no means the first time in human history that huge numbers of people will have experienced enormous irreplaceable loss *en masse*. These are exceptional levels of loss that far surpass everyday loss. Wars and genocides have caused massive loss throughout human history in death, bereavement, impoverishment, destruction, cultural vandalism and civilizational extinction. Transatlantic slavery ended the lives, worlds, comfort and identities of millions of African people and their descendants. The industrial revolution brought an end to agrarian life for hundreds of millions of people, catapulting them into urban poverty and inhumane factories and slums for the rest of their lives. The violence of all the world's empires involved similar mass loss as people were forced off their land, or relocated into changed lives of feudal monocropping, industrial labour and cultural suppression. The transition to communism in the Soviet Union and China saw massive losses through purges, famines and gulags. Plague decimated whole societies in the Middle Ages, and HIV irreparably damaged large sections of African society and ripped through the gay sub-culture of the world. Syphilis, now resurgent but curable in the US, destroyed the lives, jobs and families of millions, as did the European opium trade in China, just as opiates and fentanyl are doing in the US today.

Loss, it seems, is a characteristic of human society as it struggles destructively with political power, ideology, violence, nature, drugs and disease. None of the huge human losses in these terrible world systems and events could be replaced or truly made good. Some physical damage could be repaired and some suffering could be eased, but most loss had to be carried somehow by people who lived on and played their part in forming new societies.

Today's policies and investments that talk easily of remedy and reparation are therefore disingenuous. At best, they can support people to get beyond a bad place to a better place, and to conserve or regenerate ecosystems and culture to find new meaning in society and nature. In doing so, fairness demands that such investments are in proportion to the urgency and severity of people's condition and the greater ability of richer states to pay. The most that can honestly be expected from climate finance, loss and damage investments and adaptation programmes is that people who have lost the most can say that they did not get stuck in a terrible situation, that things did not get worse and that their lives are now on a better path. But this is recovery, not justice; it is help, not remedy or compensation.

As yet, human loss and damage from the climate emergency is nowhere near the terrible levels of loss in these other processes of human history, but it may well catch them up. Climate change could create human losses on a huge scale and, for the first time, also see the loss of the Earth as we know it and need it. This recognition rightly heightens emotions and anxiety around climate change. It is creating a sense of imminent and creeping loss as people feel and see their environments changing for the worst.

Grief for losses which have already happened is today combined with anxiety about losses that will happen soon. Human emotions around climate change are pulled between the past and the future. There is a nostalgia for what has already been lost, and an anxiety about the loss to come, for which the Australian environmentalist, Glen Albrecht, has coined the term *solastalgia*. He defines it as: "the pain or distress caused by the ongoing loss of solace and the sense of desolation connected to the present state of one's home and territory—a lived experience of negative environmental change."[3]

But alongside these feelings of loss there also exists—and there must exist—an emotional economy of determination, innovation, solidarity, cooperation and hope in which individuals, communities, companies, agencies and governments work hard to staunch losses and come through the climate emergency successfully. This is the perpetual and remarkable human capacity to keep living beyond loss and to shape a new world. It is the regenerating wave of imagination, experiment, courage, community and love that will ensure we find the knowledge and cooperation to see Earth and humanity survive. This positive emotional economy is a key place for humanitarians to be. Not only consoling but encouraging and co-creating. It is ethically important that humanitarians do not get stuck on loss and pain, but embody an ethics of transition and recovery that helps to accompany people beyond great loss.

## Loss and blame

A large part of the concern around loss and damage in the climate emergency centres on moral attribution. Whose fault is it that the whole Earth, so many humans and so much nature, are suffering from climate change? Searching honestly for the historical culprits is truly difficult and a diversion of energy away from climate action. However, finding culprits in the present is simpler and important.

Historical responsibility is easy to see in general terms but hard to pin down ethically. The cause of global warming is clearly the choice to pursue fossil fuels, eat meat and prioritize deforestation during the industrial revolution. By this measure, capitalism and centrally planned communist economies have caused most climate change. But ethically apportioning blame is genuinely difficult, because there are so many variables and too many actors to weigh in the balance of ultimate moral responsibility within such a huge world system.

Firstly, responsibility does not simply sit with Western states who originated large scale fossil fuel systems. Governments across Africa, Asia and Latin America as well as Western states opted for carbon economies; people throughout these societies personally profited (often corruptly) from them, frequently impoverishing wider society in fossil-fuelled dictatorships like Gabon or Angola. Women and men all over the world heated their houses and cooked with wood, oil, coal, charcoal and gas; wore make-up and used lotions made with petroleum products; used cars, trucks, boats, buses and planes; lit their homes and businesses with carbon-fuelled electricity; and ate a lot of meat. Most humans were greedy for carbon-based products and very few said no to them on whichever continent they lived, even when we knew about their damage from the 1990s onwards.

Then there are the positive effects of carbon to consider in the balance: its acceleration of global trade and an exceptional rise in the standard of living across the world; the medicines made, the health improved and the life expectancy raised; the schools, hospitals, universities, factories, ways of working and communal sites that were upgraded by greater energy flows; and the many inventions enabled by the carbon era, including those which will now enable us to move beyond carbon to renewables. If we are not deep ecologists who despise humans as an invasive species, then we must value these gains.

Historically, it is genuinely difficult to weigh all this and attribute moral responsibility or blame in one place by volume of the most carbon burned, not least because other people in the poorer world would clearly also have burned more if they could have done so, and many did so when they migrated to nearby cities or the Western world. Everyone was caught up in a carbon-based economy that enabled and promised so many seemingly good things.

It is much easier to detect current responsibility and culpability in two main areas. Firstly, it is morally clear that whoever is richest now—and most able to help others work through their climate-related losses—should do so. This loss adjustment is a matter of justice, but it is not specifically punitive because they have benefitted most from the carbon economy. Instead, it is simply because they have the most to share. Fairness and humanity demand that the rich should show a genuine preference for helping the poorest, and invest in positive tipping points which offer the best chance of mitigation and adaptation for the Earth. Secondly, it is much easier today to see who is wilfully continuing to burn fossil fuels, damage the environment and stall a just transition when they could be transitioning to greener options. These individuals, companies and governments should indeed be taken to court and stopped by new laws which recognize human rights and nature rights to a safe, healthy and clean environment.

## Humanitarians and loss

Loss seems more like a form of suffering that demands response than a guiding principle of humanitarian action. As such, it could be added to the principle of humanity so that humanitarian purpose becomes to "alleviate suffering *and loss* wherever it may be found." Alternatively, it could be assumed in suffering and treated more programmatically as a form of suffering to be addressed like hunger or disease.

Either way, humanitarians need to recognize the loss of nature and the specific human losses of livelihood and meaning, and draw attention to these as major consequences of climate change, a pervasive tragedy in the Earth emergency. However, the humanitarian role is not to become professional loss adjusters, who calculate and repay fair compensation for losses. Instead, the humanitarian challenge is to help people keep their life and health as they create new meaning and new good lives that honour what is lost by the making of a new society.

So, what precisely can and should humanitarians do? In wars, disasters and epidemics to date, humanitarians have never been able to make good human loss. Instead, they have been able to relieve people to some extent with lifesaving of various kinds, care and consolation. Humanitarians have also been able to avert and prevent even greater loss with early warning, preparedness, prevention and systemic support to public services in water, health and social protection. Sometimes, they have also been able to persuade powerful political forces to show restraint, and to advocate for those with bigger development budgets to invest in people's recovery. Humanitarians should continue to do all this in the climate emergency too.

A first step is to acknowledge loss as a central part of the climate emergency, and define it clearly. Environmentalists like Saleemul Huq at the IIED began this process, and humanitarians have taken it up strongly in campaigning around the COP process and operationally in their work at the community level.[4] After this, humanitarians need to be realistic about what they can help people to achieve emotionally, economically and morally.

Emotional realism is important. People can see that they are losing things or about to lose them in a changing climate, but may not want to really know it. Here, they can be helped to face the possibility of loss. As Barnett notes: "recognizing it clearly may be the best way to offset its harms and make it less existentially troubling."[5] Actively "engaging with loss" recalls the idea of positive vulnerability—that knowing one's risks is the best way to start addressing them. Planning for loss with communities enables them to emotionally prepare while thinking through desired futures and alternative options. It then becomes possible to console feelings of loss and mark its tragedy in various ways, by naming and memorializing it, while leveraging energy, expertise and social cohesion into new forms of meaning, survival and adaptation.

In this emotional work with and beyond loss, humanitarians can help people to stay positive and forward-looking. Humanitarian work should emphasize and model the virtues of fortitude, practical wisdom, ingenuity, cooperation and courage, looking for a way through. Humanitarian practice must not get stuck in simply pathologizing loss and trying to treat it. Agonizing and therapies have their place, but a certain stoicism is required to achieve the necessary pace, personal change and collective cooperation in climate action. Most communities will take a stoic approach and recognize that their own agency is their best chance of recovery and change. Humanitarians should not hold them back by defining them as traumatized or sick with loss.

Economic realism is vital too. Humanitarians must continue to work with people on their means of survival—the safety, health, income, capabilities and freedoms that can keep them alive and able to adapt. Much of this hard operational work will rightly focus on supporting people through a transition from what they have lost to new practical possibilities for achieving better life conditions in a new way of life.

Finally, humanitarians have an important role to play in the moral economy around climate-related loss by constantly bringing those who have lost most to the attention of policymakers, budget-holders and political decisionmakers at all levels. Although such large life losses can never be made good, fairness in the allocation of climate finance is possible and right. Fair allocations and just transitions should see people with the greatest loss being helped towards new economic opportunities, new skills, new landscapes and new homes so that they can play their part in sustaining themselves and a new natural environment that is safe, healthy and clean. Mobility may be an important part of such transitions and the next chapter looks at the ethics of human movement.

# 10

# Valuing Mobility

Movement is essential to survival and success in nature and humans. Life moves. Usually, we move for good reasons: to eat; to meet; to marry; to cooperate; to compete; to improve; to build; to work; to make money; to care for each other; to care for the Earth; to accumulate good things; to explore opportunities; and to find a better place. Yet very few of us are endless movers. All creatures tend to make and keep a home as well as move. We are sedentary and mobile, as are most other forms of life within the Earth community. Many species nest and migrate, roam and return, hunt and hibernate.

Climate change is a challenge to both aspects of life—moving and staying—because it is changing the environment in which we typically do both. It is difficult to remain living in the same way in areas that are becoming increasingly uninhabitable. Other areas are being recognized as better places to be because they are cooler and drier and their land is becoming more fertile, or because some societies are adapting better than others. Most animals, insects and plants are experiencing similar pressures, and many are exploring new areas and seeking out new niches. This multispecies mobility is an important sign that movement is itself a part of life on Earth. It is not as negative as the sedentary bias of national governance suggests, in which sovereignty prefers that people stay within fixed borders and nomadic people are set firm limits to their range. Movement is instead a sign of life, anticipation, experimentation and adaptation.

Movement is disruptive in ways that are both good and bad, however. It often brings opportunity, diversity and value, but it can generate hate and contest too. Human movement always creates social, ecological, economic and political change. Personal movement changes individual lives, and large collective movements change the peoples who move, the society they leave and the society they join. Because of this, movement always poses moral questions about fairness, continuity, mutualism, integration and transformation. The moral risk in human and non-human movement explains the universal ambivalence to strangers which mixes hospitality and kindness with curiosity and suspicion in a set of primal questions. Who are they? What do they want? What can they offer? What harm can they do? Are they staying or passing through? Strangers have their questions too. Is it safe? How will I survive? Will I be treated fairly or face discrimination? Will I ever make it work? Is it worth the risk?

Like loss, climate change will not be the first time in human history that there have been large collective movements relative to national and global populations. In southern Italy, it is always moving to see the stone memorials set up to remember some of the millions of Italians who migrated from Italy to other parts of Europe, the Americas and Australia in the twentieth century. Some 15 million Italians migrated between 1876 and 1920, with the total number rising to 26 million by 1976 because of terrible new poverty resulting from World War Two.[1] It is equally moving to hear the Australian accents of their visiting descendants on the beaches of Calabria and watch American-Italians come and see the church repair they helped to fund in small villages in Abruzzo, like the one in which Madonna's grandparents were born. Today, new migrant memorials are being built in Italy to remember many Africans who have died trying to cross the Mediterranean.[2]

We are not yet seeing the huge international movements of people that humanity has witnessed in the past, but we may. The largest movements so far are within states, usually as rural-urban migration. However, we can expect a lot of movement that is related in whole or in part to the climate emergency. This requires humanitarian ethics and agencies to be clear about how they value human movement alongside the equally important value of a traditional place and long-term home. Any ethics of human movement must find a balance between a sedentary bias and a movement bias, and recognize what is good and bad in both in the climate emergency.

## Movement as a human good

The recent surge in migration studies has shown how human movement within and between states is a complicated business, which is easily prone to simplistic stereotypes around its causes and effects. There are a variety of reasons why human beings move, and a variety of ways in which they move.[3]

Movement can be *seasonal* to avoid harsh weather at home by taking a regular job somewhere else for summer or winter. It can be *circular* by going to live away from home for many years, but always investing in home to return there at retirement or when savings are sufficient. It can be permanent *emigration* to another country, or *return migration* when people go back to a country of origin after emigrating. It can be *forced* movement in fear of persecution or in a forced population transfer of some kind. Movement can also be temporary as emergency *evacuation*. Finally, there is *immobility*.[4] People may be stuck or trapped, unable to move because they lack key capabilities and characteristics—like education, good health, age, ability or particular nationalities—that make them desirable abroad.

In the jargon of the day, they have very low *motility*, ability to move. But there is also voluntary immobility, where people choose to stay on their land and in their homes. Many people want to protect their ecological and cultural heritage, sometimes preferring to die with it.

There are many reasons why people move, and usually it is a combination of reasons that confirm their decision. There can be *push and pull* factors that shape a decision, such as: climate and environmental change; a sluggish economy at home and a booming one elsewhere; and reliable processes for welcoming, housing and governing migration in a booming state or area. Movement can also be driven by personal *aspirations* for better education or particular life goals, lifestyles and desires. And, of course, many people end up living somewhere else because they fall in love with someone from another place, or because their friends and social networks encourage them to come.

Each type and driver of human movement, except forced immobility, are entirely reasonable and we would want to protect them as possibilities for ourselves. Mobility offers people a chance to save or improve their lives, and to realize good things for themselves, their families and the various societies in which they move and live. All of us must surely have ancestors who moved in these ways and for these reasons, which accounts in part for why we are alive and where we live.

But largescale or slowly accumulating human movements can bring real social, political, economic and environmental challenges. Ecosystems and environments can be overwhelmed or destroyed by sudden population pressures (as often happens in forced displacement settings), rapid unplanned urbanization, or expansive commercial predation of an area's natural resources. The challenges for cultural and social cohesion are also very real as every society knows from its efforts, or refusal, to accommodate and integrate people from different places, religions and cultures. Inter-group conflict, racism and resource competition has to be worked through over decades. So, while we all value human movement in itself and know it to be an integral part of our common humanity and our common Earth, it clearly carries risks as well as benefits.

## Mobility justice and humanitarians

This puts humanitarian ethics in the climate emergency right in the middle of questions of what Professor Mimi Sheller calls mobility justice.[5] A quick glance at the world today reveals a range of "uneven mobilities" that affect individual bodies and whole groups of people differently. Various "kinetic elites" have relatively free movement around the world and access to the best sedentary spaces in which to make their homes, while others are restricted in various ways. For example, women's bodies cannot move freely around Afghanistan and are inhumanly destined to a largely sedentary and unjust life. Markers of class and colour similarly limit people's mobility, or influence how they move and self-censor when they enter certain spaces. As sovereignty is firmly protected, having often been won in blood and at great loss, certain nationalities are forbidden to cross into the territory of others.

Humanitarians are well aware of these differences in motility from long years of working in war, genocide and political repression. The ethical challenge in the climate emergency is to break down these differences and encourage a more humane, egalitarian and fairer valorization of human movement. With parts of the Earth effectively closing to large groups of people and others opening up, we need a new moral paradigm of human movement, which sees it as a necessary and good thing, which must be managed ethically and well. This is, in essence, the spirit of the non-binding *Global Compact for Safe, Orderly and Regular Migration* agreed by a majority of UN member states in Marrakesh in 2016.[6]

Climate emergency demands that we should also change the labels that we use to describe human movement. Humanitarians should do away with current humanitarian labels which international policymakers have stamped on human movement. The terms "internally displaced person" (IDP) and "migrant" have had their day, and the word "refugee" is unwisely extended to climate mobility. The label IDP has seen people stuck and rendered relatively immobile for decades because it overshadows their rightful identity as citizens in a country. Instead, it gives them an out-of-place identity, which undermines their citizenship and implies that they are the primary responsibility of the international welfare system that coined the term back in World War One and reinvigorated it in the 1990s. The migrant label has the same effect on people who have already been somewhere for generations, like Cote d'Ivoire, the Gulf States and the USA, but are then not easily integrated into those countries. Movement labels produce stigma which is, as the original Greek suggests, engraved into people's identity. This skews perceptions of their contribution and hides the wisdom and moral value of their movement.

The reaction of President Tong of Kiribati in 2008 to the suggestion that his people would be "climate refugees" showed strong resistance to mobility labels. "The people of Kiribati do not want to leave our island as 'climate refugees'", he stated firmly, "but as skilled migrants." Rejecting the refugee label, he made clear that Kiribati people wanted to be met with respect as they adapted to climate change and so granted an equal place in the world, and the agency to be a part of a world that is working to overcome climate change.[7] Today, he would probably prefer that they were not "migrants" either.

A new report by the Centre for Global Development on the international workforce needed to construct a green transition across the world makes the same point by showing that the climate emergency needs many millions of people to move.[8] All countries—and not just those in Europe and North America—will need a massive amount of people to create the "green labour", "green infrastructure" and "green jobs" necessary for climate mitigation and adaptation. People contributing to this should be highly valued as heroes of global climate action and not stigmatized with legacy labels from earlier periods of war and nationalism. The main legal focus should not be about the identity of green workers but about their climate contribution and their labour rights, with the spotlight falling on their working conditions, pay and entitlements rather than the fact that they come from somewhere else.

At the same time, mobility justice relates just as much to *not moving* as to moving. Mobility should not be over-valued as the only or best option for survival and adaptation in the climate emergency. This is especially true when many people will be moving from one high-risk place to another as the emergency intensifies. For example, moving from a struggling agricultural zone to a hazard-prone, poor and unplanned informal settlement in a rapidly expanding littoral city. This means that EU development policies which aim to help people "in place" by tackling "the root causes" of some mass movements from Africa to Europe also have value. They may not simply represent a cynical ruse to keep Africans in Africa, but instead support what most people across the Sahel and West Africa actually prefer, which is to stay where they live and make it work. A global analysis of people's preferences by the UN University shows that most people in the world "aspire to stay" rather than move.[9]

Most people have a sedentary bias and prefer home, especially if it could be economically more beneficial and better adapted to climate change. This means humanitarians should not become mesmerized by mobility but also be working hard to support people's voluntary immobility where it is feasible and preferred. Overall, however, the benefits of staying or moving will be determined by how well people, societies and nature can adapt to climate change in different places. Adaptation is, therefore, the next policy area that needs clarification in humanitarian ethics.

# 11

# Supporting Adaptation

In May 2023, State Farm, the biggest home insurance firm in the US, announced that it would be stopping all new home insurance in California. Other firms followed in California, Florida and Virginia. In many other parts of the US, the cost of home insurance has doubled and tripled. If ever there were an indicator that the climate emergency had arrived, this was it. Federal and State governments who are traditionally insurance providers of last resort are now fast becoming providers of first resort in many areas. This will not hold. In Australia, the challenge is similar and the Australian Climate Council recently announced that "Australia is fast becoming an uninsurable nation."[1]

The problem is simple. Houses and apartments are being rebuilt with insurance only to be destroyed again in the same way the following year. At the time of writing in 2024, the US has seen record numbers of storms and fires since 2020, and a huge volume of annual claims that would, until recently, have only been seen in a very exceptional year. The message from the insurance industry is clear: climate-related disasters are becoming more frequent and more intense; the cost of damage from storms, floods and fires is rising exponentially year on year; and the current business model of the insurance and reinsurance industry is breaking. In short, it makes no sense to build homes in the same way and in the same places. Humanity must adapt to a changing climate and its extreme weather which is shaping a new environment for human habitation.

## What is adaptation?

Adaptation is the second big policy of the UNFCCC process and, with mitigation, makes up the great policy duo of climate action. After a late start in the COP process, adaptation has now achieved centre stage in climate policy.[2] If we cannot stop climate change entirely by reducing emissions, then we have to adapt to it in ways which enable us to be safe, cool, dry and healthy while able to sustain the Earth, create the economy we need and enjoy the good things that make a fully human life.

According to the IPCC, adaptation is: "adjustment in natural or human systems in response to actual or expected climate stimuli or their effects, which moderates harm or exploits beneficial opportunities."[3] As such, adaptation is the process that creates the ideal of long-term resilience discussed in chapter 4. Adaptation can be *anticipatory*, and so carried out in advance, or *reactive* to what has already happened. It can be *autonomous* and *spontaneous* as conscious or unconscious action taken informally and independently by humans and nature, or it can be *planned* in highly orchestrated adaptation schemes. It can, therefore, be *private adaptation* made by individuals, families, communities and businesses, or *public adaptation* led by government.

Adaptation takes many forms in many different settings. In her excellent primer on adaptation, Professor Lisa Dale groups them into six main areas: climate services, early warning, early action and insurance; built infrastructure; nature-based solutions; agriculture, land use and food security; urban planning; migration and managed retreat.[4] However, Dale rightly points out that adaptive success is complicated to measure. The big difference between mitigation and adaption measures is their simplicity of indicators.

Gauging mitigation is relatively easy. There is really one big measure of success: reductions in greenhouse gas emissions assessed in a single indicator of global temperature. Is it going up or down in degrees Celsius? Measuring the success of adaptation is much more complicated because it involves investments in many kinds of adaptation in different places, for different types of people, and targeting different adaptation objectives. Any adaptation is also bounded by uncertainty, because initially successful adaptation could fail in five years' time when conditions change or the unprecedented happens. The medium- and long-term uncertainty of adaptation makes it a particularly hazardous task for humanitarians working with one-year budgets and near-term bias.

There are different degrees of adaptation. It can be *incremental* in a way that tweaks existing systems and ways of life to keep them going under duress. This may not change the essential precarity in which people live, but gives them a bit more survival space and so is closer to vulnerability reduction within the socio-economic status quo. For the poorest, incremental adaption is, most probably, temporarily successful coping.[5]

A much deeper form of adaptation can be *transformational*, which makes a fundamental alteration in the conditions and way of all life that is a major step change for social justice and the rights of nature. The great advantage of getting adaptation right is the so-called "triple dividend" promoted by the International Commission on Adaptation and others. Transformational adaptation avoids further losses; enables economic benefits from new investments; and produces new benefits for nature and human society by creating a better environment.[6]

Adaptation is also described as soft and hard. *Hard adaptation* refers to physical adjustments to solid structures. These may be power stations, transport, supply chains, factories, supply chains, drains, hospitals, schools and homes that make up the human built environment, and also the forests, rivers, clouds, atmosphere, seas and ecosystems that form the natural environment. Finance—as hard cash—is a fundamental component of hard adaptation. Time constitutes a hard boundary around all adaptation because time is running out in areas facing escalating ecological degradation and increasing hazards. *Soft adaptation* refers to things that are more epistemic, relational and emotional, like knowledge, skills, expertise, capability, behaviour, politics, cooperation, enthusiasm, attitude, attachment, fear, ambition and resistance.

The *adaptive capacity* of an individual, a family, a community, a species, an ecosystem, a city or a country is determined by how much hard and soft adaptation is possible for them. Successful adaptation is inevitably born of a combination of both spheres of adaptation. Each one of them can also act as barriers to adaptation. *Hard limits* to adaptation exist where change is not physically, ecologically, technically or financially possible, or there is simply not enough time in which to do them. *Soft limits* set in where adaptation might be technically and ecologically possible but is held back by insufficient political power, willingness, skills and expertise to achieve it. Levels of adaptive capacity are, therefore, key features in successful programming.

## *The ethics of adaptation*

Good adaptation—either incremental or transformational, soft or hard—is always uncertain because climate change is unprecedented and we can never be completely sure what weather, tipping points and cascading effects are coming our way. As a result, adaptation wisdom suggests we should design and implement adaptation with some key principles in mind. Back in 2009, Stephane Hallegate at the World Bank applied six classic policy strategies to adaptation, all of which still make good practical and moral sense whether one is considering crop diversification, mobility or geoengineering.[7]

First is a *no regrets* principle. This means being sure that whatever you do will produce some good in whatever situation emerges. Second is the *reversable* principle, which ensures that whatever you do can be stopped, changed or reversed, like an anticipation or early warning system that is now tracking the wrong things. *Safety margin planning* is the third, meaning that you always allow for a worse-than-expected scenario, such as building much bigger drains or cooling facilities than you currently anticipate. Fourth is to prioritize less expensive *soft solutions*. For example, change water use behaviour instead of, or as well as, making slow and costly investments in massive new water supply systems. Fifth, work to *short decision horizons* so that adaptation does not end up lagging behind changed conditions, and adaptation is redundant because it took too long. Finally, ensure *climate action synergy* in which adaptation complements simultaneous mitigation and loss and damage strategies, so that you find win-win or even dual-use solutions between adaptation and emission reduction investments, like cooling systems that run on renewable power.

Despite best planning, adaptation is haunted by the risk of *maladaptation*. Professor Lisa Schipper describes how maladaptation is much more than simply a waste of resources in a failed attempt that takes people or nature back to square one. Worse than this, maladaptation is "where exposure and sensitivity to climate change impacts are increased as a result of action taken… and an action results in conditions that are *worse* than those which the original strategies were trying to address."[8] She coins the useful term "rebounding vulnerability" to describe how adaptation bounces back on people. Like a boomerang, bad adaptation returns to hit them in another way that makes people worse off than they were originally by creating completely new vulnerabilities or deepening old ones. This might happen when farmers diversify to new crops that use less water, but then markets change and so demand for their new crops collapses. People then have to fall back on their previously failing crops or resort to high-risk urban migration. Mobility itself can be maladaptation if barriers are met along the way or the destination also faces equal climate risks or severe non-climate risks. Maladapted mobility explains the wise push-back by scholars and humanitarians against the simple hyphenated idea of "migration-as-adaptation".[9]

Equity in adaptation is another major moral challenge in the climate emergency. Financially, there is already a huge "adaptation gap" between the adaptation that needs to happen and the finance to fund it. UNEP estimates that global finance needs for adaptation are between ten and eighteen times as great than current investment in adaptation, and that one in six countries still does not have a national adaptation planning instrument to set and monitor adaptation goals.[10] So, there is a coverage problem. Worse still, many adaptation investments that are underway seem not to focus on equity as a key objective.

The first global assessment of how equity is integrated into the planning and implementation of adaptation programmes, led by Malcolm Araos, is revealing.[11] It shows that 60% of recorded projects include equity objectives and the great majority of them are in rural Africa and Asia. Adaptation in North America and Europe does not focus sufficiently on equity, and most urban adaptation has an equity blind spot. The reason Africa and Asia perform better on equity seems to be that many adaptation projects in these regions are self-organized and initiated autonomously in co-creations between communities and development organizations. This locally led developmental culture involves equity procedures in its design and specifically seeks out marginal groups.

The study concludes that "there is an urgent need to promote the integration of equity in adaptation planning and implementation." It also vindicates the policy that adaptation should be locally led and strategically planned. This commitment to locally led adaptation is set out in the *Principles for Locally Led Adaptation* agreed by the Least Developed Countries Group on climate change. Their eight principles prioritize equity considerations, low-level decision-making, and patient and predictable finance for adaptation.[12]

More than being a wicked problem, adaptation is also open to capture by wicked people. As adaptive capacity becomes a key resource, powerful and violent communities are likely to make adaptive grabs of various kinds. Siri Eriksen and her colleagues have looked at the processes of "accumulation by adaptation that widens inequality and undermines broader adaptation goals".[13] Power can become unfairly concentrated in key natural sources of adaptation like rare earth, water and temperate land, or in key adaptive technologies that could enable broad-based positive tipping points if more fairly shared. In competition over adaptation, there is also the risk of adaptation conflicts and the use of violence to co-opt and secure adaptation. The risk of violence will surely rise as adaptation becomes more valuable and strategic, and if equity is not factored into local, national, regional and global adaptation schemes.

## Humanitarians and adaptation

Adaptation is essential in human and nature's response to climate change, as humanity and nature both seek out long-term resilience. As such, adaptation becomes a moral imperative in the humanitarian mission and the new moral paradigm set out in chapters 5–8. Like State Farm at the top of this chapter, humanitarians will increasingly face the impossibility of repairing things as they were, or doing business as they did it before. In many situations, it will make no moral sense to help people stand up again in the same way or in the same place, simply to be knocked straight down again. Such humanitarian action would constitute repeated failure at best and humanitarian negligence at worst.

Adaptation should, therefore, rightly become a guiding principle of humanitarian ethics. The 2021 *Climate and Environment Charter for Humanitarian Organizations* affirms this in its very first commitment to "helping people to adapt to the impacts of the climate and environment crisis".[14] But how, and how much? As we have seen, adaptation is a massive whole-world challenge of technical, social, ecological and political change that requires budgets and skills way beyond humanitarian aid. Successful transformational adaptation is the ideal for humanity and nature. However, the extreme settings, limited budgets and short timeframes in which humanitarians operate make incremental adaptations of various kinds the best they are likely to achieve with people in extremis.

Humanitarian adaptation can clearly include early warning, anticipatory action and whatever hard and soft adaptation measures humanitarians can support people to take to improve their conditions of life, ecosystems and livelihoods in places where they are. It should also include humanitarian support to new patterns of mobility which make good moral sense. These may be spontaneous adaptive movements that need intermediate humanitarian accompaniment along the way, or initial humanitarian support in the preparation and start-up of more strategically planned processes of managed retreat and relocation away from unadaptable climate and environments.[15]

In carrying out this highly pragmatic and inevitably limited form of humanitarian adaptation, humanitarians should be bound by three other ethical commitments in all their adaptation efforts.

Firstly, humanitarians should champion equity in adaptation. Impartiality requires this and they are often well placed to spotlight, reach and correct inequity vertically and horizontally. Humanitarian aid can start vertical programmes with overlooked communities of humans and nature to make some of form of adaptation real for them. In the process, they can also widen the areas of the Earth community that are brought to the attention of larger adaptation programmes by linking previously marginalized landscapes into government and commercial schemes. This simultaneously vertical and horizontal expansion of inclusion is something that humanitarians have already done especially well, by deepening and widening social protection schemes and people's financial inclusion through cash programming in times of crisis.

Secondly, to lean into greater equity, humanitarians should always work with processes of locally led adaptation and connect these initiatives with wider strategic planning on adaptation by the development system.

Thirdly, humanitarians, like everyone else, must do their best to manage and evade the constant shadow of maladaptation. In doing so, they should plan and work with as much foresight as possible for humanity and nature today and future generations of the Earth community. Here, it makes sense to be guided by Hallegate's sound precautionary principles of no regrets, reversable measures, safety margins, and a preference for soft solutions, short decision cycles and synergy with mitigation measures. Not all of these will be feasible in every situation, but they can still act as useful guides and clear evidence of well-intentioned adaptive deliberation by a humanitarian team.

As humanitarians increasingly accept and operationalize the moral imperative of climate change adaptation as part of their core purpose, they will gradually shape a new practice of humanitarian adaptation that will ensure that incremental adaptation, at least, becomes possible for more people and more ecosystems. This will be a valuable contribution to a more equitable expansion of adaptation as the climate emergency escalates, and may keep people going while larger transformational adaptation deepens and expands to reach them.

# 12

# Humanitarianism 2.0

In this book I have argued that the humanitarian ethics and principles we have inherited from 1965 must be significantly updated in a new moral paradigm that takes much greater account of nature and the future. This new ethical framework represents a humanitarian revolution demanded by the Earth emergency, which requires humanitarians to look beyond human suffering alone, and beyond the needy present to the needs of future life as well. This is "Humanitarianism 2.0"—a new version of humanitarian ethics to match the challenge of the Earth emergency.

In essence, this new paradigm reflects a moral deepening of humanity's sense of identity, life and time. The Earth emergency has rightly deepened our sense of identity with the Earth. Its potential assault on all life and its rapid environmental change has reawakened a sense of mutualism between humanity and nature as our source of life. This rightly renews our shared identity with the Earth community as well as with our unique human community.

The new "we" we find in common with all other life on Earth means that humanitarianism is not only for humans. It is also for nature for which we feel compassion and responsibility in a new all-life ethic. With time running out, and new technical ability to anticipate the near and medium term, the future also imposes itself ethically upon humanitarians to demand that what we do in the present also makes moral sense in the future. We cannot think only about the emergency of today. The emergencies that we can now foresee in six months, two years or in the next generation rightly demand humanitarian actions in the present as precautionary, anticipatory and intergenerational aid. This sees a clear extension in our temporal morality.

Revising our core principles in line with this new humanitarian consciousness is a major change, and will be seen as idealistic and over-ambitious by some humanitarians. However, in many ways, my argument is simply making more explicit an ethical evolution which is already well underway in humanitarian ethics. Humanitarian work, codes, standards and charters have routinely expressed ethical and practical concern for the environment and for the sustainability of relief interventions in recent decades. Environmental ethics have influenced humanitarians for some time. Today, our understanding of the intensity of the Earth emergency makes us intellectually and operationally ready to state these new commitments more clearly at the heart of humanitarians ethics. My revisions in this book, therefore, explicitly affirm earlier incremental ecological moves and promote them from subsidiary ideas of humanitarian good practice to the level of fundamental principles that fit the Earth emergency we face.

As the foundational expression of humanitarian values and purpose, our beloved principles of humanity and impartiality should now be revised to make each of them more ecological and impartiality more future-facing. I hope my suggested revisions in chapters 6 and 7 are a useful start to this process. Existing operational goals of reducing vulnerability and increasing the resilience of humanity and nature should become core principles of operational purpose. So too should precaution, which includes anticipation and adaptation, and so consolidates the future's place in humanitarianism's operational ethics.

At the heart of the new moral paradigm that I am suggesting in this book are several specific ethical and operational shifts. These are all set out in the previous chapters, but it may help to express them more succinctly in this final chapter. Each one of them requires ethical, attitudinal and institutional change in humanitarians today—a new humanitarian mindset.

## A new doctrine of humanity

Chapters 5 and 6 have argued that we need to revise our sense of being human and so reframe our understanding of humanity itself. We need a new doctrine of humanity for this moment in the twenty-first century, which helps us to move from a human-centric Anthropocene era to the more inclusive Ecocene era proposed by Mihnea Tănăsescu. Ecocene ethics specifically prompts us to shape a universal politics with and for the whole Earth community, not just for humans. In the Earth emergency, it makes no sense to treat humans as a singular species that can be cared for in isolation because, in reality, we can only live by virtue of our relationships with other species and with the air, wind, rain, warmth, sea, ice, soil and rock that combine to give us life. Humanity is unique but is not alone nor self-sufficient. We must, therefore, recognize our shared being and our mutual life as part of the wider Earth community. We do not live only as a human community, and we cannot live only as a human community. Protecting and respecting human life is necessarily about protecting and respecting a diverse multitude of other life which also has its own significant and beautiful place in the world. This makes the humanitarian mission an inevitably ecological mission.

This recognition of shared ontology and shared politics with nature requires a new anthropology of humanitarian action that recognizes humanity as part of the Earth community and understands that it is impossible to help humans survive without also helping the wider Earth community to survive. The idea that humans are Earthlings reminds us of this truth. It guides our humanitarian action as Earth action too in all kinds of programmes supporting ecosystems, food security, risk reduction, species conservation, One Health policies and nature-based solutions.

At the heart of this new doctrine of humanity is a deep appreciation of what is valuable, beautiful and lovable in the world around us, and so a new commitment to show humanity to plants, animals, insects and to the natural world of water, land, air and rock. Just as humanitarians have increasingly come to recognize human rights in the first years of the twenty-first century to consolidate a universal sense of the human community, so we will also come to appreciate the rights of nature in the next part of the century and so consolidate the Earth community. Gradually, industrial humanity is rediscovering more ancient and traditional worldviews of shared being, shared space and common life with nature. Humanitarians have an important part to play in protecting and building this exciting new sense of Earth community.

## A new landscape approach to need

This more ecological doctrine of humanity, which sees humans more integrated within the Earth community, leads naturally to a new way of seeing suffering and need. The assessment of need and the response to need must look holistically at humans and nature simultaneously if we are to value the survival of both and the life-giving mutualism between humanity and nature. This is best done in a landscape approach which looks across all life in a specific area and not at human life alone. This is implicitly done already in some food, health and livelihood programming which looks ecologically at need and response. However, the humanitarian system itself must use landscapes more deliberately as the main framework of humanitarian need as it responds to drought, heatwaves, floods, storms and wars.

Today's calculation of humanitarian need is driven and designed around the estimated suffering of individual humans. The accumulated number of "people in need" (PINs) is the measure of human need in a district or a country and forms the basis of a humanitarian appeal. But this basic unit of analysis in humanitarian action needs to change in the Earth emergency. Instead of people in need, PINs should refer to "places in need" and humanitarian assessments should gauge the significance, suffering, vulnerability and resilience of all life across a given landscape so that assessments take account of human and ecological need in a response which targets the Earth community as a whole.

This shift in mindset may seem especially alarming to some of today's humanitarians who see so much human suffering around them and have learned to focus firmly on the preciousness of every human life. But, if we are honest, we know that the best way to save human life in disaster and war is by sustaining a wide variety of other life that produces light, air, water, shade, food, income, health and beauty.

Even the most desperate peri-urban IDP camp or bombed out urban street is an all-life landscape, which is brought back to life by recognizing the importance of nature and humanity together. And each one of these human communities will need the rivers, fields, trees and ecosystems of its wider landscape to support humanity. In this way, a landscape approach need not marginalize humans but include them more fully in life-supporting environments and systems. Nor does it exclude individual care to people in discreet human programming around healthcare, social work, education and livelihood generation. Indeed, each of these can often have important nature-based dimensions.

## A new type of humanitarian organization

A landscape approach requires a new type of humanitarian organization that is a mixture of humanitarian and ecological expertise. Many of today's humanitarian and development institutions are challenged by Thomas Hale's "institutional lag" as their mandates, structure and capacities lag behind the times. For example, most organizations are siloed into either human or environmental mandates and skills, and most humanitarian organizations have spent the last thirty years specializing in war humanitarianism rather than climate-related disaster. However, the Earth emergency and the Earth community require merged institutions that can combine human and nature programming.

To be fit for purpose for the Earth emergency, humanitarian organizations urgently need to join forces with ecological agencies in institutional mergers or landscape-based partnerships which see them working as one. The Earth emergency needs integrated agencies, not parallel lines of human and ecological programming. Emergency financing also needs to be quickly transformed so that a single budget is able to support simultaneous human and ecological assessment and response.

Landscape thinking and response is also likely to be the optimal approach to massive needs arising from climate-related disasters. The sheer volume of human, plant, animal and ecosystem needs will be so large in some situations that individualizing need may prove too difficult. A humanitarian focus on the human and environmental systems that support all life may be the best way to support survival when a detailed focus on individual human needs becomes impractical.

## *Future-led thinking*

A determination to think more about the future and to think more *from* the future is the next mindset change required by humanitarians in the Earth emergency. This becomes essential for three main reasons. Firstly, because we can, and because precaution is a binding principle in the climate emergency. Today, we can anticipate more than we could. As seen in chapter 8, in certain situations we can see weeks and months ahead, and this foresight demands action from us in the present as anticipatory aid or adaptation for current and future generations. Secondly, if we engage seriously with landscapes in integrated human and ecological organizations, then we will need to think long-term as we invest in adaptation and nature-based solutions which take years to mature. Thirdly, much of the climate emergency will be unprecedented and cascading. This means only certain things in the present and the past will be good guides to humanitarian action. Instead, we need to be thinking prospectively of extraordinary things that might happen, but have never happened yet, as we plan and respond today.

Getting past our short-term bias and our constant looking backwards will be hard, but it will be helped in new merged organizations in which humanitarians work with ecologists who are trained to think in deeper systems and longer timespans. Essentially, humanitarianism in the Earth emergency requires humanitarians to think more proactively and consequentially than we usually do.

## A new sense of ecological justice

The humanitarian paradigm suggested in this book shares the new sense of ecological justice that is rising around the world today. This is perhaps most simply expressed in the "3Is" model of justice proposed by Joyeeta Gupta and others, which recognizes three significant arenas of justice.[1]

*Interspecies justice* recognizes the importance of justice between species and the natural world to sustain Earth system stability for all life. This is the balance of justice between humans and creatures, and between humans and natural formations like air, sea, soil, rock and forest. For humanitarians, this is the arena of the environment and the importance of focusing on reduced operational emissions, disaster risk reduction, biodiversity conservation, and adaptation working with nature-based solutions.

*Intragenerational justice* focuses on fairness between countries, communities and people in this generation. This represents humanitarian concerns with the distributive justice of aid between humans in access to safety, food, health, water, education and livelihoods. It is the challenge of fairness and impartiality across all levels of human society in the climate emergency, and so relates to urgent need in climate-related disasters, legacies of loss and damage, mobility justice, and just transition and adaptation.

*Intergenerational justice* is fairness between generations. This sees the current generation looking ahead to the needs of future generations of humanity and nature, and also backwards to past generations whose emissions and pollution have damaged the Earth and human communities—whether knowingly or not. For humanitarians, this means constructing humanitarian norms and laws that protect future generations, and designing DRR, anticipation and adaptation which works for the future life of the Earth community. It also means saving lives today—human, plant, animal, insect, river and sea life—to ensure reproduction and the new life of a next generation. It also means representing and supporting communities of humans and nature who have lost most from past actions, helping them on a path to the future that is founded on fairness, recovery and resilience.

Each one of these arenas of justice requires *procedural justice*—good processes in which fair policies are agreed and distribution decisions are effectively enacted. This will involve humanitarians in the development of new laws and practices to protect humans and nature in the climate emergency. Legal ways will need to be found to bring the life of other species into humanitarian consideration alongside human life. Humanitarians must become involved in the emerging construction of human rights to a safe, healthy and clean environment that is underway at the UN, and the corresponding rights of nature which will see the natural world increasingly recognized and respected in the norms, laws and institutions of human politics. Humanitarians must start representing nature as well as people in political and legal arenas, and bring the needs and the agency of nature to the table in humanitarian decision-making.

As governments and international society struggle to agree on these new arenas and procedures of justice, humanitarian NGOs should start to assert them in their own institutional mandates and practices. In so doing, they will be quicker than the COP and other UN processes and can pioneer important political innovations that states later adopt. Humanitarian NGOs have done this many times before in their role as norm entrepreneurs by creating new norms and practices, which then cascade through international society. The history of war humanitarianism is one long story of NGO innovation influencing states and international law.[2] The challenge now is to do the same thing in the climate emergency.

## Cooperation, courage and hope

The Earth emergency demands particular virtues from humanitarians as well as new thinking and practice. The next decades will be difficult ones for humanity, as they so often are. The challenge we face now is not just the improvement of our life on Earth under increasingly difficult circumstances but also the protection of the Earth itself. This responsibility can feel daunting and even triggering. However, as usual, we are not alone. We are many. Humanity is blessed with the mind of the hive and with some outstanding individuals. So, we can cooperate— and we must cooperate—with other people and other organizations as we bring conviction, energy, insight, science, experimentation and collaboration to the challenge.

Cooperation means all the usual stuff about breaking down siloes and thinking across our disciplines and organizations to find good partnerships. The humanitarian sector is very small in comparison with governments, business, climate finance, development and academic organizations engaged in the climate emergency. But this means that we can be agile and connect with them all. Particularly important will be strong cooperation and mergers with ecological institutions and with major adaptation initiatives. As we have seen, it is unlikely that humanitarians will be able to offer much more than incremental adaptation, so we must link people and nature well to much deeper systems of positive and transformational adaptation that are being driven by business or government.

Cooperation in this new ethical framework of humanitarianism also means cooperating with nature. This is the essential idea behind nature-based solutions, but we must elevate cooperation with nature to a clear principle of action. Ecological humanitarianism requires that we work with nature just as we work with people. This means developing the idea of nature's participation and empowerment in humanitarian programming as a subject and agent in humanitarian success. Humanitarians must be talking about what nature needs and what nature can do, just as much as they talk about what people need and what people can do. And, of course, they must be talking about what humans and nature can do together. In the process, this will extend the idea of "mutual aid" from a practice between people to a practice between nature and people.

The Earth emergency requires courage. It can feel rightly terrifying at times. Not surprisingly, the word terrifying has its roots in the Latin word *terra*, meaning earth. This shows how the Earth has always been a source of fear for us as well as joy. Humanity has been terrified for as long as we have been terrestrial, and it is not unusual to be afraid of what the Earth can do to us as it quakes, storms, floods and burns. It is part of the human condition to be terrified. Yet humanity has courage to deal with this, and we must relish and share our courage as we work to save and help all life in this latest terrifying episode on Earth. We can find courage in ourselves and from others around us. Above all, we can find courage in our love of the Earth. This is where we live and where we are from, so we must struggle to keep it.

In the process of this struggle, humanitarian work can be brutal and tough, because the suffering of people and nature is so awful and tragic. Hope helps us here. Hope ensures that, although we may sometimes feel sad, we should never feel doomed. Philosophers are often sceptical of hope and suspicious of it as a delusion. They think it best to dispense with hope, because agreeing that there is none in the current crisis is the best place from which to start thinking new things to get us out of this mess.[3] I disagree. I like hope and I enjoy hoping. It gets me up in the morning and it keeps me going. My hope is not a mindless optimism, more a commitment to the future and a certain resolve to try and find it despite the trials of the present. Like courage, I tend to gather hope from people and places around me, and from within me and my Catholic faith. But every humanitarian worker must decide whether they think and work better when feeling doomed or feeling hopeful. The main thing is to be motivated by one or the other.

## A revolutionary moment

A couple of people who have heard me talk about this book have wondered why we need to change our fundamental principles of humanity and impartiality and adopt a whole new ethical framework for humanitarianism today. Surely, they said, we could just add on an extra principle about the importance of the environment and nature, and then work to that as well. But this does not work for me.

It is not enough to simply add a new ecological principle to our existing list of humanitarian principles. Our change in humanitarian perspective today must involve a completely new all-life view of humanity and the Earth together, and an explicit commitment to show humanity and compassion to the Earth community and cooperate with it. An additional principle on the environment will always remain subservient to the original principles as an add-on. Nature and the future will never get the integral place they require in our new era of Earth emergency, and the minds of humanitarians will stay biased towards human life alone. Instead, we need to be bold. This is a time for humanitarian revolution and not for humanitarian tweaking.

# Notes

## 1.  UPDATING HUMANITARIANISM

1. Report of the Twentieth International Red Cross Conference in the *International Review of the Red Cross*, 1966.
2. Marina Sharpe, 'It's All Relative: The Origins, Legal Character and Normative Content of the Humanitarian Principles', *International Review of the Red Cross*, 2023.
3. The Code of Conduct for the International Red Cross and Red Crescent Movement and NGOs in Disaster Relief, accessed at https://www.ifrc.org/sites/default/files/2021-07/code-of-conduct-movement-ngos-english.pdf
4. The Sphere Project and Core Humanitarian Standard, accessed at https://spherestandards.org/
5. The Climate and Environment Charter for Humanitarian Organizations, accessed at https://www.climate-charter.org/
6. IFRC and Climate Centre, *A Guide to Climate-Smart Programmes and Humanitarian Operations*, Geneva, 2023.
7. The English word "environment" comes from the French word "environ", meaning that which is around us.
8. For an excellent short history of environmental ethics, see: Jason Kawall, 'A History of Environmental Ethics', in Stephen Gardiner and Allen Thompson (eds), *The Oxford Handbook of Environmental Ethics*, Oxford University Press, Oxford, 2015, chapter 2.

9. Declaration of the UN Conference on the Human Environment, accessed at https://documents-dds-ny.un.org/doc/UNDOC/GEN/NL7/300/05/PDF/NL730005.pdf?OpenElement

10. Brundtland Report, published as *Our Common Future*, Oxford University Press, Oxford, 1987.

11. See the *Common Declaration on Environmental Ethics* by Pope John Paul II and the Ecumenical Patriarch His Holiness Bartholomew I, made at Venice on 10 June 2002.

2. AN EARTH EMERGENCY

1. Miki Khanh Doan et al, 'Counting People Exposed to, Vulnerable to, or at High Risk from Climate Shocks', Policy Research Working Paper 10619, World Bank, Climate Change Group, Washington DC, November 2023.

2. ND-GAIN, accessed at https://gain.nd.edu/about/

3. Bryan Harris, 'Brazil Rolls Out Dengue Vaccines as Cases Rise Sharply', *Financial Times*, 25 February 2024.

4. Holmes Rolston also uses the term "earthlings" in his *A New Environmental Ethics: The Next Millennium for Life on Earth*, Routledge, London, 2020, chapter 7.

5. Johan Rockstrum, Joyeeta Gupta et al, 'Safe and Just Earth System Boundaries', *Nature*, Vol 619, 6 July 2023.

6. Tim Lenton et al, *Global Tipping Points Summary Report*, University of Exeter and Bezos Earth Fund, December 2023, accessed at https://global-tipping-points.org/

7. The suggestion of Gaia as the name and image for Lovelock's Earth system came from his friend William Golding, the Nobel Prize-winning novelist.

8. James Lovelock, *We Belong to Gaia*, Penguin, London, 2006, p. 2.

9. Ibid, p. 17.

10. Ibid, p. 5.

11. For a useful discussion of this trend, see: Regan Burles,

'Another Geopolitics? International Relations and the Boundaries of World Order', *International Studies Review*, Vol 23 (4), December 2021.

12. Mihnea Tănăsescu, *Ecocene Politics*, Open Book Publishers, Cambridge, UK, 2022, pp. 10–11.


3.   EARTH'S ELEMENTS AS HUMAN HAZARDS

1.   Alice Bell, *Our Biggest Experiment: A History of the Climate Crisis*, Bloomsbury, London, 2021, p. 9ff.

2.   Simon Clark, *Firmament: The Hidden Science of Weather, Climate Change and the Air That Surrounds Us*, Hodder and Stoughton, London, 2022, pp. 60–64.

3.   Zeke Hausfather, 'Methane and Other Gases', in Greta Thunberg, *The Climate Book*, Penguin, London, 2021, chapter 2.3.

4.   Ibid.

5.   Clark, op. cit, pp. 173–178.

6.   Matthieu Auzanneau, *Oil, Power and War: A Dark History*, Chelsea Green, White River Junction, 2018, p. XIII.

7.   Ibid, p. IX.

8.   Ibid, p. 2.

9.   Ibid, p. 35, citing Ron Chernow, *Titan: The Life of John D. Rockefeller Sr*, Vintage, New York, 2004, p. 248.

10.  Kate Crowley et al, 'Climate and Weather Hazards and Hazard Drivers', in eds Kelman, Mercer and Gaillard, *The Routledge Handbook of Disaster Risk Reduction Including Climate Change Adaptation*, Routledge, Abingdon, 2020, chapter 5, pp. 35–46.

11.  The IPCC used the term "radiative forcing" to describe the changes and perturbations in the atmosphere brought about by human agency, but the term "climate forcing" is now more routinely used.

12.  Clark, op. cit, chapter 7, pp. 41–54.

13.  Stefan Rahmstorf, 'Warming Oceans and Rising Seas', in Greta

Thunberg, *The Climate Book*, Penguin, London, 2021, chapter 2.11.

14. Michael Le Page, 'Hurricanes Are Becoming So Strong We May Need a New Scale to Rate Them', *New Scientist*, 5 February 2024.

15. World Meteorological Organization, *Provisional State of the Global Climate 2023*, Geneva, 16 November 2023, p. 1.

16. Ibid, p. 1.

17. Copernicus, '2023 Is Warmest Year on Record', accessed at https://climate.copernicus.eu/global-climate-highlights-2023#:~:text=2023%20marks%20the%20first%20time,than%202%C2%B0C%20warmer

18. Centre of Disease Control (CDC), 'Extreme Heat Can Impact Our Health in Many Ways', accessed at https://www.cdc.gov/climateandhealth/pubs/EXTREME-HEAT-Final_508.pdf

19. Yuta J. Masuda et al, 'Impact of Warming on Outdoor Worker Well-Being in the Tropics and Adaptation Options', *One Earth*, Vol 7 (3), 15 March 2024, pp. 382–400

20. Xiao Shi et al, 'Spatiotemporal Variations in the Urban Heat Islands across the Coastal Cities in the Yangtze River Delta, China', in *Marine Geodesy*, Vol 44 (5), 2021, pp. 467–484.

21. World Meteorological Organization 2023, op. cit, p. 1.

22. Sourav Mukherjee et al, 'Increase in Compound Drought and Heatwaves in a Warming World', *Geophysical Research Letters*, 8 December 2021.

23. Qi Zhao et al, 'Global, Regional and National Burden of Mortality Associated with Non-Optimal Ambient Temperatures from 2000 to 2019: A Three-Stage Modelling Study', *The Lancet*, July 2021; see also OCHA and IFRC, *Extreme Heat Report*, Geneva, 2022.

24. Trocaire, *Still Feeling the Heat: How Climate Change Continues to Drive Extreme Heat in the Developing World*, Maynooth, 2017.

25.　Yan He et al, 'Substantial Increase of Compound Droughts and Heatwaves in Wheat Growing Seasons Worldwide', *International Journal of Climatology*, 28 December 2021.

26.　Kendra K. McLauchlan et al, 'Fire as a Fundamental Ecological Process—Research Advances and Frontiers', *Journal of Ecology*, February 2020.

27.　UNEP, *Spreading Like Wildfire: The Rising Threat of Extraordinary Landscape Fires*, Nairobi, chapters 3 and 4, pp. 55–80.

28.　OECD, *Taming Wildfires in the Context of Climate Change*, Paris, p. 12.

29.　Matthias M. Boer and Víctor Resco de Dios, 'Unprecedented Burn Area of Australian Mega Forest Fires', *Nature Climate Change*, Vol 10, March 2020, pp. 170–172.

30.　Ibid, p. 172.

31.　Davide Faranda, 'The Great Cold Debate: Is Climate Change to Blame for Extreme Cold Spells?', *EGU Blog*, 27 December 2022, accessed at https://blogs.egu.eu/divisions/np/2022/12/27/extreme-cold-spells/

32.　Helle Abelvik-Lawson, 'Cold Weather and Climate Change Explained', Greenpeace, 22 March 2021, accessed at https://www.greenpeace.org.uk/news/cold-weather-and-climate-change-explained/#:~:text=Sudden%20freezing%20cold%20isn't,don't%20usually%20experience%20them

33.　For example, UN OCHA's 2023 winterization programme in Ukraine, accessed at https://www.unocha.org/publications/report/ukraine/ukraine-winter-response-plan-october-2023-march-2024-issued-august-2023

34.　Zhao et al, op. cit.

35.　Gianluca Pescaroli and David Alexander, 'Understanding Compound, Interconnected, Interacting and Cascading Risks: A Holistic Framework', *Risk Analysis*, Vol 38 (11), June 2018.

4.   PRECAUTION, VULNERABILITY AND RESILIENCE

1.   For comparison, the 2022 floods in Pakistan covered 75,000 km$^2$ according to UNOSAT.

2.   Chris Courtney, *The Nature of Disaster in China: The 1931 Yangzi River Flood*, Cambridge University Press, Cambridge, 2018, pp. 3–5.

3.   The following sections draw on Courtney (ibid), especially his chapter 4, pp. 121–152.

4.   *Sendai Framework for Disaster Risk Reduction*, accessed at https://www.undrr.org/implementing-sendai-framework/what-sendai-framework

5.   Kathryn Jean Edgerton-Tarpley, 'From Nourish the People to Sacrifice for the Nation: Chinese Responses to Disaster in Late Imperial and Modern China', *Journal for Asian Studies*, Vol 73 (2), May 2014.

6.   For a good quick sketch of the evolution of DRR, see: Rajarshi DasGuspta and Rajib Shaw, 'Disaster Risk Reduction: A Critical Approach', in eds Kelman, Mercer and Gaillard, *The Routledge Handbook of Disaster Risk Reduction Including Climate Change Adaptation*, Routledge, Abingdon, 2017, chapter 3, pp. 12–23.

7.   Piers Blaikie, Terry Cannon, Ian Davis and Ben Wisner, *At Risk: Natural Hazards, People's Vulnerability and Disasters*, Routledge, London, 1994, p. 23 and p. 160.

8.   Timothy O'Riordan and Andrew Jordan, 'The Precautionary Principle in Contemporary Environmental Politics', *Environmental Values*, Vol 5, 1995, pp. 191–212.

9.   See the excellent short summary of these different positions in Simon Caney, 'Climate Change', in eds Moellendorf and Widdows, *The Routledge Handbook of Global Ethics*, Routledge, Abingdon, 2015, chapter 27, especially pp. 376–380.

10.  Henry Shue, 'Deadly Delays, Saving Opportunities', in eds Gardiner et al, *Climate Ethics: Essential Readings*, Oxford University Press, Oxford, chapter 8.

11. IFRC, *Climate-Smart Disaster Risk Reduction*, Geneva, 2020; and Aktion Deutschland Hilft, *Enhancing Efficiency in Humanitarian Action through Reducing Risk: A Study on Cost-Benefit of Disaster Risk Reduction*, Bonn, 2021.

12. Catriona Mackenzie, Wendy Rogers, and Susan Dodds, *Vulnerability: New Essays in Ethics and Feminist Philosophy*, Oxford University Press, New York, 2014.

13. Hugo Slim, *Humanitarian Ethics: The Morality of Aid in War and Disaster*, Hurst, London, 2015, pp. 213–216.

14. Kate Brown, 'Vulnerability: Handle with Care', *Ethics and Social Welfare*, Vol 5 (3), September 2011.

15. Errin C. Gilson, *The Ethics of Vulnerability: A Feminist Analysis of Social Life and Practice*, Routledge, Abingdon, 2014.

16. Sofia Morberg Jamterud, 'Acknowledging Vulnerability in Ethics of Palliative Care—A Feminist Ethics Approach', *Nursing Ethics*, Vol 29 (4), 2022, pp. 952–961.

17. Gilson, op. cit, p23.

18. Marina Berzins McCoy, *Wounded Heroes: Vulnerability as Virtue in Ancient Greek Literature and Philosophy*, Oxford University Press, Oxford, 2013, p. 210.

19. Joachim Boldt, 'The Concept of Vulnerability in Medical Ethics and Philosophy', *Philosophy, Ethics and Humanities in Medicine*, Vol 14 (6), 2019.

20. Stefan Hallegate et al, *Lifelines: The Resilient Infrastructure Opportunity*, World Bank, Washington DC, chapter 4.

21. Michelle Ward et al, 'Impact of Mega-Fires on Australian Fauna and Habitat', *Nature Ecology and Evolution*, Vol 4, 20 July 2020, pp. 1321–26.

22. World Wildlife Fund, *Australia's 2019–2020 Bushfires: the Wildlife Toll*, 2020, accessed at https://assets.wwf.org.au/image/upload/v1/website-media/resources/Animals_Impacted_Interim_Report_24072020_final?_a=ATO2Ba20

23. Yvonne Walz et al, 'Disaster-Related Losses of Ecosystems and Their Services. Why and How Do Losses Matter for Disaster Risk Reduction?', *International Journal of Disaster Risk Reduction*, Vol 63, July 2021.

24. Nathalie Doswald et al, 'Ecosystems' Role in Bridging Disaster Risk Reduction and Climate Change Adaptation', in Kelman et al, op. cit, chapter 12, pp. 116–128.

25. Thea Hilhorst, 'Classical Humanitarianism and Resilience Humanitarianism: Making Sense of Two Brands of Humanitarian Action', *Journal of Humanitarian Action 3*, Article 15, 2018.

26. Jeong Han Kim et al, 'Resilience from a Virtue Perspective', *Rehabilitation Counselling Bulletin*, Vol 61 (4), 2018.

27. Carl Folke, 'Resilience', *Ecology and Society*, Vol 21 (4), December 2016.

28. IFRC, *Framework for Community Resilience*, Geneva, 2020, p. 6.

29. Peter Rogers et al, 'Resilience and Values: Global Perspectives on the Values and Worldviews Underpinning the Resilience Concept', *Political Geography*, Vol 83, November 2020.

5. HUMANITY AND NATURE

1. For a full discussion of the principle of humanity as it stands, see Hugo Slim, *Humanitarian Ethics: The Morality of Aid in War and Disaster*, Hurst and Oxford, 2015, pp. 44–55.

2. Reema Chopra, *Shifting Paradigm: Towards a Transformative and Holistic Vision of Humanity*, Master's Thesis, Institute of International Cooperation in Education, University of Zug, 9 February 2020.

3. Ibid, p. iii.

4. Nick Lane, *The Vital Question: Why Is Life the Way It Is?* Profile Books, London, 2016, p. 8.

5. Jakob von Uexkull, *A Foray into the World of Animals and*

*Humans—With A Theory of Meaning*, University of Minnesota Press, Minneapolis, 2010. It must be noted that von Uexkull had clear Nazi sympathies.

6.  Angela S. Stoeger, 'Elephant Sonic and Infrasonic Sound Production, Perception and Processing', in eds Cheryl S. Rosenfeld and Frauke Hoffmann, *Neuroendocrine Regulation of Animal Vocalization*, Elsevier, 2021, chapter 12, pp. 189–199.

7.  *The Holy Quran*, Sura 6, Livestock, Aya 38, translated by M.A.S Abdel Haleem, Oxford University Press, Oxford, 2004, p. 82.

8.  Robin Wall Kimmerer, *The Democracy of Species*, Penguin, London, 2013, p. 1.

9.  Ibid, p. 18.

10. For an excellent overview of different sacred theologies of nature, see: Karen Armstrong, *Sacred Nature: How We Can Recover Our Bond with the Natural World*, The Bodley Head, London, 2022.

11. Thomas Aquinas, *Summa Theologica*, Part One, Question 8.

12. *The Upanishads*, translated by Juan Mascaro, from the Chandogya Upanishad, Penguin, London, 1965, p. 114.

13. Workineh Kelbessa, 'African Worldviews, Biodiversity Conservation and Sustainable Development', *Environmental Values*, Vol 31 (5), December 2021.

14. Vandana Shiva, *Earth Democracy: Justice, Sustainability and Peace*, Zed Books, London, second edition 2016.

15. *Universal Declaration of the Rights of Mother Earth*, accessed at https://www.rightsofmotherearth.com/declaration

16. Mihnea Tănăsescu, Ecocene Politics, Open Book Publishers, Cambridge, UK, 2022, pp. 23–26.

17. Pope Francis, *Laudato Si'*, paragraphs 48 and 49.

18. Jaboury Ghazoul, *Ecology: A Very Short Introduction*, Oxford University Press, Oxford, 2020, pp. 20–24.

19. Kelbessa, op. cit, p. 583.

20. Anna Tsing, 'Unruly Edges: Mushrooms as Companion Species', *Environmental Humanities*, Vol 1, 2012, pp. 141–154.

21.  A word brought into mainstream biology by Lynn Margulis in her pioneering work on symbiosis.

22.  I am grateful to my uncle Jalil for telling me this Hadith, versions of which can also be found here: https://www. myquranstudy.com/articles/Signs-002-Tree-Stump-Crying. aspx

6.    DEEPENING HUMANITY

1.    Bruce Mazlish, *The Idea of Humanity in a Global Era*, Palgrave Macmillan, New York, 2009.

2.    Here I am drawing on the ecological theology of Thomas Berry; see Mary Evelyn Tucker and John Grim eds, *Thomas Berry, Selected Writings on the Earth Community*, Orbis, Maryknoll, 2014, especially chapter 3.

3.    Xi Jinping, 'Green Development Model and Green Way of Life', in *The Governance of China*, Vol II, p. 428.

4.    *The Epic of Gilgamesh*, translated by Andrew George, Penguin, London, 1991, Tablet XI, pp. 88–95; The Bible, Genesis, Chapters 6–8.

5.    Hugo Slim, 'The Power of Humanity: On Being Human Now and in the Future', *ICRC Law and Policy Blog*, 30 July 2019, accessed at https://blogs.icrc.org/law-and-policy/2019/07/30/ power-of-humanity-being-human-now-future/

6.    Mary Midgley, *Animals and Why They Matter*, Georgia University Press, Athens, 1983, p. 14.

7.    *Book of Genesis*, Chapter 1 verse 27 and Psalm 8 verse 5.

8.    Celia Deane-Drummond, *A Primer in Ecotheology: Theology for a Fragile Earth*, Cascade Books, Eugene, 2017, p. 29.

9.    Bruno La Tour, *Down to Earth: Politics in the New Climate Regime*, Polity, Cambridge, 2017.

10.   Mihnea Tănăsescu, *Ecocene Politics*, Open Book Publishers, Cambridge, UK, 2022, pp. 161 and 156.

11.   Ibid, p. 157.

12. Midgley, op. cit, chapter 10.
13. This section draws on Mihnea Tănăsescu, 'Representation, Democracy and the Ecological Age', *Quaderns de Filosofia*, Vol X (2), 2023, pp. 121–132.
14. Midgley, op. cit, pp. 83–88.
15. Mihnea Tănăsescu et al, 'Rights of Nature and Rivers in Ecuador's Constitutional Court', *The International Journal of Human Rights*, 7 February 2024.
16. Ibid, p. 3.
17. See Midgley's discussion of this point about animals, op. cit, p. 48.
18. David Hume, *An Enquiry Concerning the Principles of Morals*, ed. Tom L. Beauchamp, Oxford University Press, Oxford, 1998, p. 19.
19. Pope Francis, *Laudato Si'*, paragraph 89.

7.    EXTENDING IMPARTIALITY

1.    For an analysis of the traditional humanitarian interpretation of impartiality, see: Hugo Slim, *Humanitarian Ethics: The Morality of Aid in War and Disaster*, Hurst, London, 2015, pp. 56–64.
2.    David Wallace-Wells, 'The Uninhabitable Earth', *The New York Magazine*, 9 July 2017.
3.    Red Cross Red Crescent Climate Centre, *Regional-Level Climate Factsheet: Middle East*, 2021.
4.    Hugo Slim, *Humanitarians and the Climate Emergency: The Ethical, Practice and Cultural Challenge*, GPPi Berlin, 28 June 2023, accessed at https://gppi.net/2023/06/28/humanitarians-and-the-climate-emergency
5.    Miki Khanh Doan et al, 'Counting People Exposed to, Vulnerable to, or at High Risk from Climate Shocks', Policy Research Working Paper 10619, World Bank, Climate Change Group, Washington DC, November 2023.

6. For a fuller explanation of this retreat to survival needs, see Hugo Slim, *How Should We Define and Prioritize Humanitarian Need?*, Norwegian Centre for Humanitarian Studies, Oslo, 13 November 2023, accessed at https://www.humanitarianstudies.no/resource/how-should-we-define-and-prioritise-humanitarian-need/

7. *Catechism of the Catholic Church*, Geoffrey Chapman, London, 1994, paragraph 2447.

8. Joel Glasman, *Humanitarianism and the Quanitification of Human Needs: Minimal Humanity*, Routledge, Abingdon, 2020, p. 260.

9. I am grateful to Liana Ghukasyan at IFRC and Meg Sattler at Ground Truth for thinking me through this reversal in my discussions with them on humanitarian need.

10. IFRC and WWF, *Working With Nature: How Nature-Based Solutions Reduce Climate Change and Weather-Related Disasters*, June 2022.

11. Ibid, chapter 4.

12. Blaise Pascal, *Human Happiness*, translated by A.J. Krallsheimer, Penguin, London, 1966, Pensée Number 47, pp. 9–10.

13. Thomas Hale, *Long Problems: Climate Change and the Challenge of Governing Across Time*, Princeton University Press, Princeton, 2024.

14. Ibid, p. 14 and p. 13.

15. Pope Francis, *Laudato Si'*, paragraph 159.

16. Simon Caney, 'Justice and Posterity', in eds Ravi Kanbur and Henry Shue, *Climate Justice: Integrating Economics and Philosophy*, Oxford University Press, Oxford, 2019, pp. 157–174.

17. Stephen M. Gardiner, 'A Perfect Moral Storm: Climate Change, Intergenerational Ethics and the Problem of Moral Corruption', in eds Gardiner et al, *Climate Ethics: Essential Readings*, Oxford University Press, Oxford, 2010, pp. 87–98.

18. Matthew Rendell, 'Discounting and the Paradox of the Indefinitely Postponed Splurge', in eds Ravi Kanbur and Henry Shue, *Climate Justice: Integrating Economics and Philosophy*, Oxford University Press, Oxford, 2019, pp. 175–187.

19. Hale, op. cit, chapter 3.

20. Ibid, chapter 4.

21. Ibid, chapter 5.

22. Vaclav Smil, *How the World Really Works: A Scientist's Guide to Our Past, Present and Future*, Penguin, London, chapter 2, pp. 45–75.

23. Jean Pictet, *The Fundamental Principles of the Red Cross: Commentary*, Henry Dunant Institute, Geneva 1979, pp. 12–13.

## 8.  EMBRACING ANTICIPATION

1. Pope Francis, *Laudato Si'*, paragraph 36.

2. Dave MacLeod and Sarah Klassen, *A Practical Guide to Seasonal Forecasts*, Braced and Shear, London, June 2019.

3. Chris Funk et al, 'Tailored Forecasts Can Predict Extreme Climate Informing Proactive Interventions in East Africa', *Earth's Future*, Vol 11, 15 May 2023.

4. UK Met Office and Red Cross Red Crescent Climate Centre, *The Future of Forecasting: Impact-based Forecasting for Early Action*, Geneva, 2020, p. 13.

5. United Nations, *Early Warning for All Initiative*, quoting the Global Status Report 2022 and the Global Commission on Adaptation, accessed at https://www.un.org/en/climatechange/early-warnings-for-all

6. German Red Cross et al, *Anticipatory Action in 2022: A Global Overview*, Berlin, 2023.

7. World Food Programme, *Scaling Up Anticipatory Action for Food Security*, WFP, Rome, April 2023.

8. World Food Programme, *Climate Risk Financing, Anticipatory*

*and Early Actions for Climate Hazards*, WFP, Rome, November 2022.

9. The START Fund, *Preparing for Climate Change Impacts on Small-Medium Scale Crises*, London, 2023.

10. Mihai Nadin, 'Anticipation: An Annotated Bibliography', *International Journal of General Systems*, Vol 39 (1), 2010.

11. Roberto Poli, 'Anticipation: What about turning the humanities and social sciences upside down?', *Futures*, Vol 64, 2014, pp. 15–18.

12. My summary of Rosen is drawn from A.H. Louie, 'Robert Rosen's Anticipatory Systems', *Foresight*, Vol 12 (3), 2010, pp. 18–29.

13. See the section on temporality in Husserl in the *Stanford Encyclopaedia of Philosophy*, accessed at https://plato.stanford.edu/entries/self-consciousness-phenomenological/#TempLimiReflSelfCons

14. I am grateful to Mihnea Tănăsescu's work for introducing me to Whitehead's thought.

15. For a longer discussion of dependency as an ethical risk see, Hugo Slim, *Humanitarian Ethics: The Morality of Aid in War and Disaster*, Hurst, London, 2015, pp. 97–98.

16. See the *Code of Conduct* at https://www.ifrc.org/our-promise/do-good/code-conduct-movement-ngos; and the *Humanitarian Charter* at https://spherestandards.org/humanitarian-standards/humanitarian-charter/

## 9. RECOGNIZING LOSS

1. Maximilian Kotz, Anders Levermann and Leonie Wenz, 'The Economic Commitment of Climate Change', *Nature*, Vol 628, pp. 551–557, 17 April 2024.

2. Jon Barnett et al, 'A Science of Loss', *Nature Climate Change*, Vol 6, November 2016.

3.    Glenn A. Albrecht, *Earth Emotions: New Words for a New World*, Cornell University Press, Ithaca, 2019, p. 38.
4.    See for example, Red Cross Climate Centre and Flood Resilience Alliance, *Key Findings Related to Loss and Damage*, Geneva, 2023; Caritas, *Unheard, Uncharted: A Holistic Vision for Addressing Loss and Damage*, November 2023; Paul Knox-Clark and Debbie Hillier, *Addressing Loss and Damage: Insights from the Humanitarian Sector*, Flood Resilience Alliance, May 2023.
5.    Barnett et al, op. cit, p. 978.

## 10.  VALUING MOBILITY

1.    Stephen Castles, Hans De Haas and Mark J. Miller, *The Age of Migration, International Population Movements in the Modern World*, Fifth Edition, Palgrave Macmillan, Basingstoke, 2014, p. 93; Donna R Gabaccia, *Italy's Many Diasporas*, Routledge, London, 2003.
2.    'Migrant Crisis: How Lampedusa Memorial Reached British Museum', BBC News, 21 January 2016, accessed at https://www.bbc.co.uk/news/world-europe-35360682
3.    This section follows the excellent summary in Alex de Sherbinin et al, 'Migration Theory in Climate Mobility Research', *Frontiers in Climate*, 10 May 2022.
4.    For a good overview of immobility, see Caroline Zickgraf, 'Theorizing (Im)mobility in the Face of Environmental Change', *Regional Environmental Change*, Vol 21 (4), December 2021.
5.    Mimi Sheller, 'Theorising Mobility Justice', *Tempo Social: Revista de Sociologia da USP*, Vol 30(2), 2018; and her book *Mobility Justice: The Politics of Movement in an Age of Extremes*, Verso, London and New York, 2018.
6.    *The Global Compact on Safe, Orderly and Regular Migration*, accessed at https://www.iom.int/resources/global-compact-safe-orderly-and-regular-migration/res/73/195

7. Carol Farbotko and Heather Lazrus, 'The First Climate Refugees? Contesting Global Narratives of Climate Change in Tuvalu', *Global Environmental Change*, Vol 22, 2012, p. 383.

8. Sam Huckstep and Helen Dempster, *Meeting Skill Needs for the Global Green Transition: A Role for Labour Migration?*, Centre for Global Development, CGD Policy Paper 318, Washington, 2024.

9. Alix Debray, Ilse Ruyssen and Kerilyn Schewel, *Voluntary Immobility: A Global Analysis of Staying Preferences*, UNU-CRIS Working Paper Series, No 8, December 2022.

11. SUPPORTING ADAPTATION

1. Ian Smith, Attracta Mooney and Aime Williams, 'The Uninsurable World: What Climate Change Is Costing Homeowners', *Financial Times*, 13 February 2024.

2. Lisa Schipper, 'Conceptual History of Adaptation in the UNFCCC Process', *Review of European Community and International Environmental Law*, May 2006.

3. *IPCC Glossary of Terms 2012*, accessed at https://archive.ipcc.ch/pdf/special-reports/srex/SREX-Annex_Glossary.pdf

4. Lisa Dale, *Climate Change Adaptation*, The Earth Institute, Columbia University Press, New York, 2022.

5. E. Lisa F. Schipper, 'Maladaptation: When Adaptation to Climate Change Goes Very Wrong', *One Earth*, Vol 3(4), October 2020, p. 410.

6. Global Commission on Climate Adaptation, *Act Now: A Global Call for Leadership on Climate Resilience*, 2019.

7. Stephane Hallegate, 'Strategies to Adapt to an Uncertain Climate Change', *Global Environmental Change*, Vol 19(2), 2009, pp. 240–247.

8. Schipper, op. cit, p. 409.

9. Patrick Sakdapolrak, Marion Borderon and Harald Sterly, 'The

Limits of Migration as Adaptation', *Climate and Development*, Vol 16(2), 2024, pp. 87–96.

10. UNEP, *Underfinanced. Underprepared. Adaptation Gap Report 2023*, Geneva.

11. Malcom Araos et al, 'Equity in Adaptation-Related Responses: A Systematic Global Overview', *One Earth*, Vol 4, 2021, pp. 1454–1467.

12. Least Developed Countries Group, *Principles for locally led adaptation*, 2021, accessed at https://www.iied.org/principles-for-locally-led-adaptation

13. Siri Eriksen et al, 'Adaptation Interventions and Their Effect on Vulnerability in Developing Countries: Help, Hindrance or Irrelevance?', *World Development*, Vol 141 (4), May 2021.

14. *The Climate and Environment Charter for Humanitarian Organizations*, accessed at https://www.climate-charter.org/

15. *Planned Relocation in the Context of Disasters and Climate Change: A Guide for Asia-Pacific National Societies*, IFRC, Geneva, 2021.

## 12.  HUMANITARIANISM 2.0

1. Joyeeta Gupta et al, 'Earth System Justice Needed to Identify and Live within Earth System Boundaries', *Nature Sustainability*, Vol 6 (6), March 2023, pp. 630–638.

2. Hugo Slim, *Solferino 21: War, Civilians and Humanitarians in the Twenty First Century*, Hurst and Company, London, chapter 5.

3. For example, Mihnea Tănăsescu starts his book *Ecocene Politics* from this hopeless position. For a full treatment of hope, see Terry Eagleton, *Hope Without Optimism*, Yale University Press, 2015.

# Index